STRENGTHENING TECHNOLOGICAL CAPABILITY

Strengthening Technological Capability

Strategy for Developing Countries

Editors
Pradosh Nath
S. Visalakshi

GYAN PUBLISHING HOUSE
New Delhi-110002
in Collaberation with
NATIONAL INSTITUTE OF SCIENCE TECHNOLOGY & DEVELOPMENT STUDIES (NISTADS)
New Delhi-110 012

STRENGTHENING TECHNOLOGICAL CAPABILITY

Industry, Technology managment

ISBN: 81-212-0640-5

Published in 1999 in India by
Gyan Publishing House,
5, Ansari Road, Darya Ganj
New Delhi-110002

Laser Typesetting at Elite Art, New Delhi
Printed at Chawla Offset Printers, Delhi

Contents

Preface

In the literature on technology and development in the last two decades, one observes that terms like "technology transfer," "facilitators of technology transfer", intra-country and inter-country problems in technology transfers, have been slowly and steadily replaced by "university-industry interactions," "inter-firm alliances," "locating Research and Development (R&D) outside the countries," "networked organizations," "virtual organizations", etc. These developments can be traced to mainly three types of influences: namely globalization, new technologies like information technology and biotechnology becoming promising commercial propositions and the diminishing support of financial resources of the State to public sector R&D.

These developments have significant implications for countries like India. There is an increasing awareness that the competitiveness is through effective technologies rather than better resource endowment in terms of natural resources or cheaper human skills. Under these circumstances a look at the organizations involved in R&D is worthwhile to examine how they relate to industrial and national wealth accumulation. These organizations of R&D came into being with certain objectives and functioning modes under an ethos which allowed them to work almost autonomously—without necessarily networking with the non-R&D actors.

While we see evidence of quick adaptation to the changing circumstances in industrialized and some of the newly industrialized countries, its manifestation in India requires a careful look. What surfaces in the changed scenario is the emphasis of the research organizations serving the markets effectively and efficiently. It is necessary to evaluate the present set-up in the context of the recent changes in demands on them. This should clarify the possible modalities of making the R&D organizations more responsive to the demands of user markets as well as of high science-based technologies.

The workshop, the proceedings of which form the basis of this book, had deliberated on the above issue. It is apparent from the papers, that instead of "competition" as a key concept for acquiring market orientation, the function of key concept is "cooperation".

There are papers which give the experiences of other countries like those of the Republic of Korea and Brazil which had tackled similar requirements through their strategies. These provide a reference point. There are papers on Indian experiences where environment for such cooperation has been attempted by the State. Some papers deal with specific experiences of industries and research institutions in working towards establishment of such inter-dependent complementary arrangements.

I am happy that NISTADS' group on Management and Economics of Technical Innovation (METI) along with the Federation of Indian Chamber of Commerce and Industries (FICCI) has not only brought to the fore the issues involved but also sensitized industry, R&D organizations, and representatives of the Government. I am confident that this volume is timely and would be useful in initiating purposeful interaction among the concerned partners, namely, R&D organizations, industry and the Government.

Ashok Jain

Editors' Note

This book is the outcome of the workshop on **Institutionalization of R&D Industry Partnership** held in April, 1996. The workshop has been sponsored jointly by FICCI and NISTADS. Dr Ashok Jain, Director, NISTADS, Dr Amit Mitra, Secretary-General, FICCI extended their support and encouragement towards the success of the workshop. Dr Aditya Trivedi of FICCI kept the basic spirit of the workshop alive by ensuring wide participation of industrialists from different parts of the country. The intense discussions which went through during the course of this workshop form the basis of this volume.

We thank the participants from industries and the academic and research institutions from India and abroad for their contribution during the workshop and in the present volume.

Geeta Kathuria provided the much needed secretarial and logistic support. Anil Koli completed the tedious job of formatting and layout of the text. We thank both.

The book has been conceived immediately after the workshop. Soliciting from the authors dispersed over different parts of the country and abroad and having light work schedules had delayed the publication much beyond the editors' initial expectations.

While the editors take the responsibility for ensuring the thematic coherence and broad content of the contributions, the individual authors remain responsible for the facts, figures and information presented in their articles.

Editors

List of Contributors

Araoz Alberto	Fellow, Fletcher School of Law and Diplomacy, Tuft University, Medford, MA, USA.
Banerjee, P.	Scientist, National Institute of Science Technology and Development Studies, New Delhi.
Chaudhury, Shekhar	Professor, Indian Institute of Management, Ahmedabad,
Jamal, Tabassum	Scientist, National Institute of Science Technology and Development Studies, New Delhi.
Lee Jang, Jae	Senior Researcher, Science Technology Policy Institute, Seoul, Korea.
Machado, F.M.	Director, Technology Services, UNIDO, Geneva.
Mani, Sunil	Associate Fellow, Centre for Development Studies, Trivandrum.
Moitra, S.D.	Professor, Indian Institute of Management, Calcutta.
Mrinalini, N	Scientist, National Institute of Science Technology and Development Studies, New Delhi.
Rao, N. J	Director, Central Pulp and Paper Research Institute, Saharanpur.

Roy, Santanu	Scientist, National Institute of Science Technology and Development Studies, New Delhi.
Sandhya, G.D.	Scientist, National Institute of Science Technology and Development Studies, New Delhi.
Virmani, B.R.	Professor, Administrative Staff College of India, Hyderabad.
Visalakshi, S.	Scientist, National Institute of Science Technology and Development Studies, New Delhi.
Yim, Deok Soon	Senior Researcher, Science Technology Policy Institute, Seoul, South Korea.

Introduction

Reorganization of R&D: The Institutional Imperatives

Publicly-funded Research and Development (R&D) laboratories played a crucial role in development in all industrialized countries. Such laboratories in European countries and the USA have been forerunners in industrial R&D. Japan has rediscovered the need of public investment in R&D for technological development of SSEs to launch fully public-funded 195-odd prefecture laboratories. Newly industrialized countries like South Korea, Taiwan and Singapore have used government-promoted and funded R&D laboratories for achieving strategic industrial and technological aspirations of the respective countries.

India can claim the distinction, among developing countries, of having the largest and oldest established chain of publicly-funded R&D laboratories. But unlike many other countries, the setting up of such laboratories in India preceded the emergence of a strong industrial base which could fruitfully utilize the capabilities existing in those laboratories. In the case of South Korea, Singapore and Taiwan, and for that matter Japan, public investment in R&D has always been strategic; keeping pace with the strategic interest of growing industrialization.

These historical facts on records indicate:

(a) Public-funded R&D has a crucial role to play in the process of industrialization and technological progress of a country; and

(b) There is a need of strategic alliance between industrial R&D and industrialization.

Having realized this, the task is to look for suitable mechanism, scope and opportunities of forging strategic alliance between R&D setups and Indian industries. The importance of this task is required to be re-examined in the context of the thrust towards globalization and liberalization. The critical issues, therefore, are: (a) meaning, if any, of publicly funded R&D in India in the age of globalization; (b) articulation of an interdependent relationship between R&D and industries; (c) driving force of long-term partnership and (d) organizational imperatives for forging a partnership. In what follows, the above issues are critically examined in separate sections.

The Context of Globalization

Has globalization made R&D (particularly public funded R&D) in developing countries redundant? Will global sourcing of technology and R&D gradually replace domestic R&D laboratories' setup as a wasteful expenditure? Industrialized countries' superiority in this regard is unquestionable. Is it, therefore, economically meaningful to engage scarce resources for undertaking R&D? Furthermore, what Galbraith said decades ago is more valid today that industrial R&D has become so expensive that it is beyond the means of even the largest corporates of developing countries. Besides these, there are numerous flippant statements about the performance of public R&D in India. Anybody from industry at any point of time would dish out 100 odd cases against public funded R&D in India. Of course, there is no dearth of similar views against industries from R&D organizations.

A more critical understanding of the issue depends on how we interpret globalization and what our strategies are at a firm's level, industry's level and national level for reaping the benefit of globalization. We propose to understand globalization as an opportunity for coming out of the economic, political and social vulnerability of our nation.

Does globalization mean making best quality goods available in the Indian market for the producers and consumers of India? Making home market available to the rest of the world is definitely half of the story. Globalization has to be seen as international bench-marks for products and production activities. It can, therefore, as well mean manufacturing products of international standard and possibly setting up new bench-marks.

Let us take the concrete example of the passenger car industry of Japan. Did Japan bank only on receiving the state-of-the-art technology from the West to make the car industry of Europe and the USA run for Japanese technology? Was this great achievement possible only by sourcing technology from the successful firms in the West? Classic examples are cases of Toyota and Nissan. Nissan entered into three years' collaboration with the British company Austin. It geared up its R&D centre to indigenize and learn the technology during the short period of collaboration. The collaboration was not extended beyond the period of initial agreement. Nissan made an all-out effort to get the best out of the collaboration within that period. Toyota did not have any collaboration with any international giants of that period. But it engaged both Kyoto and Tokyo university scientists and engineers to undertake R&D on special steel, engine and other critical components related to motor car before rolling out a single passenger car from its factory. The rest is history.

The point of distinction is, of course, the long-term technological aspiration and related strategy at the firm level and national level. Peck and Goto studied the trajectories of development in the technological strength of Japan. The most profound conclusion they arrived at was that technology was not like water which flows naturally from a higher level to a lower level. To achieve technological competitiveness a long-term technological strategy is to be built up having R&D at its core. R&D strength is necessary for selecting the best technology, for learning it and also for making a technology work. All these questions are irrelevant if the objective is to have the world's finest products ready-made in India. We know that in the long run this is self-defeating since to be able to buy something one has to be able to sell something equivalent, or otherwise the nation's future is an imperilled sovereignty.

In a globalized economy it is not very difficult to attract a reasonable foreign investment, or source the best available technology from the global market by providing suitable industrial environment. But to be at the frontier of industrial technology is a different ball game. The flow of foreign investment and foreign technology to any level of satisfaction does not ensure the country any position of pride in the world of fast changing technology. On the contrary, the same may result in gradual erosion of our entrepreneurial capability, industrial and technological knowledge bases.

Being at the frontier of technology means developing capabilities in identified frontier areas. This is what R&D is all about. It will not be out of place here to mention a misplaced perception of industries about R&D. Among many allegations against the Council of Scientific and Industrial Research (CSIR), the oft-repeated one is that CSIR laboratories mostly provide bench level technology and never or rarely a fully packaged one. We forget that CSIR is an organization of R&D laboratories. Nowhere in the world does an R&D laboratory provide a fully packaged technology. What generally industries refer to is technology available from another company. That technology is always tested, tried and packaged and not as one from an R & D laboratory.

The state and direction of R&D in a particular field of technology define the frontier of that technology. Being technology recipient firms and industries, these and the country in general are deprived of any clue to the ever-expanding frontier of technology and the recipient becomes a perpetual laggard.

The question is most pertinent in the Indian context. India, unlike any other developing country, has a long history of high standard entrepreneurial capability. It can also boast of highly skilled technical and scientific manpower, the credibility of which is unquestionable in the whole range of technological ventures. All these have the potential to make the country emerge as a world leader. The question is whether we want such a global status for ourselves or make ourselves dependent on the largesse of the developed world? In a globalized scenario, making our presence marked and enhancing our bargaining position in the global market would mean developing a long-term strategy on the basis of critically assessing our strengths and weaknesses. We have a strong feeling that in such a strategy the industry has to play a more pro-active role to rope in both R&D institutions and government.

R&D institutions, without exception, do not have any intrinsic driving force either for formulating such a strategy or for taking initiatives. They have to be utilized. If financial resources are the constraint for the highly expensive R&D in frontier areas, the already established basic infrastructure available with a long chain of publicly funded R&D laboratories provides the material basis for a partnership between Indian entrepreneurs and R&D systems. In the absence of it, we shall be kept watching while our entrepreneurs will become subservient to big companies from different corners of the globe and our scientific and technical skills will be utilized for the long-term interest of the same companies.

R&D Industry: Partnership is Intrinsic

Industrial R&D can briefly be described as a process of scientific confirmation of the technological solution of a problem articulated in the production system. Once this possibility is confirmed, there come the series of activities called developmental technologies to be finally introduced in the production system. In a section of literature in this field such a relation has been postulated to be market relationship; R&D laboratories are producers of knowledge, and industries are users of knowledge. One can, therefore, see R&D expenditure as investment for production of knowledge to be sold as a commodity in the market. The rest, therefore, can be straitjacketed to textbook economics of consumer and producer behaviour, and the price is determined through hypothetical auction where one tries to squeeze out the other. That is market, and if it is perfectly competitive, everybody is happy with the right price.

Even without getting into any intricate detail it is not very difficult to see the incompatibility of such a model in describing the process of development of a technology from R&D to innovation stage. The existence of a market mechanism for R&D is not as conspicuous as it is in the case of any other commodity like a machine or any other consumer goods. Furthermore, the price mechanism of the market works on the assumption of full knowledge. The basis of R&D-industry transaction, on the contrary, is absence of knowledge the generation of which is a joint product of both R&D and production system with a well defined primary role for each. A closer analogy is to a team in a football game where team members feed each other with the ball to move together into the territory of the other team.

It is, therefore, more of partnership than rivalry. In fact, Schumpeter, an economist who elaborately dealt with the issue of technological innovation, concluded that for technological innovation perfect competition neither ever existed, nor was it desirable. Another celebrity, Nobel laureate K.J. Arrow proved with a formal model that a market kind of approach with perfect competition probably can explain minor innovation activities of firms. Arrow firmly concluded that for major innovations institutional effort is essential. What we are suggesting is that it is the institution of partnership between industry and R&D which can make the technological aspirations of developing countries like India a reality.

Partnership: The Driving Force

Partnership is best understood with the analogy of games. In the game of soccer involving a team of 11 players, each has a primary responsibility assigned to him. Although each has a separate responsibility, they have a common end to achieve. Again all players are interdependently related to each other with their individual responsibility. The common end of a team in a game is to win against the opponent. A game cannot be won if team members, the partners, have the aim of relying solely on individual skills instead of developing a network of complementary skills for all other team members. Again, a network of complementary skills for all partners does not build up automatically even if the end is common. The coach of the team plays a crucial role in defining the strategy in consultation and agreement with all team members. Strategies are essentially chalked out by a fair measure of strengths and weaknesses of the team members and also those of the members of the opponent team.

Coming back to the partnership of our concern, the R&D and production systems have their own defined role to play. The question is to win over what? The game is, we suggest, catching up with the rest of the world, to win technological superiority over other global players. But this is to be strategized by the government who has to play the crucial role of a coach. In this game, industry and R&D cannot play by their own rules. If they do, it would be finally failure for both.

A tripartite strategy is to be formulated for a synchronized network between the major players. It is to be noted that like the coach, the government also does not take direct part in the game, and instead restricts itself to the formulation of strategy and facilitating skill development of the team members, through a legal, monetary and fiscal framework.

The analogy of soccer probably ends there. The problem arises when there are many players in an industry. And it is indeed so in reality. Industry is a group of firms. Each firm may have independent aspirations which may be in conflict with other firms in the same industry. In such a situation what does partnership between industry and R&D mean?

To our understanding the partnership is really not between one firm in an industry with R&D laboratories. Individual firms on the basis of their own strategy would forge varieties of relationship with R&D laboratories. Industry -R&D partnership is to be understood in terms of the common interest of an industry or a group of industries. This is possible if the focus of the partnership is shifted from industry- specificity to technology specificity. Most of the modern technological areas, like micro-electronics, biotechnology, new materials, etc., cut across the industry boundary to find use in all and sundry industries. There could be two kinds of R&D activities: (a) R&D on strategic areas of frontier technology and (b) R&D on industry-specific technology. The partnership for the latter kind of R&D may be left to the individual firms in an industry. In fact, that would be a partnership between the in-house R&D of a firm with the R&D laboratories.

The former is, however, a partnership of different kinds where many industries can join hands together for developing capabilities in a generic technology area having future industrial use. Capabilities built up in such areas can be used by firms according to their long-term strategies.

Another benefit of this is the creation of highly skilled human resources which would meet industrial need for skilled manpower in development and application of new technologies in their respective areas.

The driving force of such partnership is the long-term national interest by creating a national pool of capabilities in strategically defined areas. This national pool of skill is for the benefit of industries and individual firms for their long-term technological and market aspiration. As we have mentioned earlier in a globalized scenario, the technological and market aspirations have to have the globe as a reference point and not the domestic market alone. And any dent in the global market can be made only by developing and suitably using the highly skilled human resources.

The benefit for the R&D institution, although not so tangible, is quite substantial. The main aspiration of an R&D institution and R&D personnel is to be in the frontier of knowledge and to get recognition for their skill. They really want to see their skill being used and the horizon of it getting continuously expanded.

A strategic design to make a dent in selected areas in the global market, therefore, ensures benefit for both the partners. The mutual gain is the driving force of such a partnership.

Partnership: The Organizational Imperatives

What is the suitable organizational design which can uphold the spirit of partnership for achieving long-term strategic technological aspiration of industries as well as the nation and of course the R&D institutions? A recently concluded study on the best practices of R&D institutions has suggested that a definite way of achieving the partnerships is by allowing a more active role for users of R&D results in the decision-making processes of research institutions. There has to be a way of reflection of industries' aspirations and expectations from research laboratories.

Here we are proposing a 3-tier R&D system with a specific role of government for addressing three broadly different types of industrial needs.

To avoid prolonged digression we are not trying to provide an elaborate discussion on the existing arrangements. We shall refer to them only for the sake of building up our argument.

The first tier could be for the small scale enterprises related to regional development. The need for regional development cannot be overemphasized. Even highly industrialized countries of Europe as also Japan have undertaken programmes for revitalizing technology support systems for regional developments. The basic understanding is that a country may be highly industrialized when its regions suffer from lack of development and related economic and social disparities with grave political implications.

In India, we have an already existing organizational infrastructure for providing technological and related supports to Small Scale Enterprises (SSEs) in general and those in specific regions. At the apex level there are bodies like Small Industry Development Organization (SIDO), National Small Industries Corporation (NSIC), etc., with their arms in state levels. The lowest administrative structure is District Industries Centre (DIC) at the district level. There are three basic problems with the DIC kind of structures: (a) resource, (b) strategic planning and (c) regional commitment.

The resource problem arises because of the centralized nature of the administration of DICs. Although in terms of geographical location it is decentralized, administrative control is centralized through complex arrangements with the central, state and district level administrations. Strikingly, the industries in the district have no role to play about the services DICs have to provide to the local industries.

DIC's activities in no way reflect the existence of any strategic planning of a region or a district. The basic assumption behind its functioning is that the user of its services would approach the respective DIC for the specific needs. This, we feel, is quite a strong assumption where the role should be the creation of a group of industrial entrepreneurs. Again resource mobilization (both human and financial) capability of DICs is in no way related to the potentiality of the district or the region. It is the capability of the apex body which decides the activities of DICs.

It is possible to visualize DICs playing a more important role with greater autonomy. DICs can be seen as a trust promoted by government and run by local administration and industries. The trust can draw with the help of selected experts the strategy of the local development of industries and technical inputs required for that. The same trust can be entrusted to set up laboratories with the autonomy of sourcing funds, recruitment of suitable manpower, and making laboratories accountable to the trust for its activities and achievements.

The main thrust here is participation of local industries, formulation of a long-term strategy for regional development, and accountability of the laboratories.

The second tier is the constitution of industry-specific R&D laboratories. As we have argued above, to be at the frontier of industrial R&D, resource requirements could be well beyond the capability of any firm in an industry. A Cess Fund approach is undoubtedly a superior approach to the problem of resource mobilization for industry-specific R&D laboratories. In such cases R&D laboratories should be under the ownership of a specific industry association which would formulate the strategy and direction of R&D activities of such laboratories and funded by Cess Fund with suitable contribution from the government. As discussed earlier, major R&D activities in such cases would be targeted to generic research types for capability building. It can also undertake the development of common facilities for the use of industries, particularly in the areas of standards and quality assurance. In India there are R&D laboratories owned by industry associations with mixed success. It is not difficult to list their failures. We would still argue that those failures do not really indicate the inefficiency of the unique organizational structure of such laboratories. In fact, these laboratories fulfil all basic conditions, namely, accountability, industries participation and resource mobilization efficiency, to become a successful case of organizational design.

Failures in these cases, we feel, are because of the prevailing industrial environment which did not have the global best as the bench-mark. Also, even within an industrial association framework in most of the cases, neither the association nor individual members nursed and nurtured these laboratories as their own baby. Industries with a general lack of focus on R&D considered these laboratories as more of a burden than a source of technological strength in strategic areas. In a changed environment the attitude is likely to change.

At the third tier are the laboratories with a broad national mandate. These are the laboratories which have to reorient themselves for strategic technology targeting. Technology targeting as opposed to industry targeting has to be understood in terms of frontier areas of technologies the uses of which are not industry-specific but open up possibilities for all types of industries. Concrete examples are micro-electronics, biotechnology, new materials, etc. Areas of national capabilities in such cases have to be chosen in accordance with the industrial thrust of the country, industries' perception about future technological needs and necessary manpower development for that purpose. The most difficult process in such cases is to ensure industries' response and participation. The best mode is to create technology-specific committees with representatives from associations of potential user industries. There have to be changes in the internal organization of R&D laboratories.

In many cases R&D institutions have been seen as having research divisions in the line of traditional disciplines of science and technology. And in many cases these divisions are rigid over the lifetime of R&D laboratories. In some cases a few new divisions are added here and there while core divisions remain as they were in the beginning.

Technology targeting has two kinds of implication on organizational design; (a) the divisions have to be technology based instead of a rigid academic discipline based; and (b) with the changes in technological scenario, and technological needs, groups have to be as much flexible as to adjust to the new and emerging needs.

In fact, no division can be formed a priori. If we visualize an active role for the industries in decision making of priority technology areas of R&D, formation of such a research group and not division has only to be a simultaneous process.

This should not be considered as a concrete proposal for reorganization. It is only indicative of initiating a discussion. The spirit of reorganization suggested is to ensure for the industry, i.e., the user of R&D, its participation in a proactive way in the activities of R&D laboratories. What we have tried to argue is the need for a partnership tuned to a strategy for long-term benefit and catching up with the rest of the world. We have also argued that in such a partnership industry has to have a more active role and probably the role of an initiator to play.

The Present Volume

The issues raised above and the views expressed in the workshop are not new or novel. Rather it is some sort of a consolidation of the various views expressed by researchers, policy-makers and other concerned. It has been strongly felt that given this understanding issues related to strategies for technological capability building for economies like India's should be discussed in a forum where industries, R&D institutions and government could freely exchange their views. With this objective in mind a two-day workshop had been organized in April, 1996, by the Management and Economics of Technological Innovations (METI)— a research group of NISTADS jointly with FICCI (Federation of Indian Chambers of Commerce and Industries). FICCI's participation ensured industries' participation. Both METI and FICCI together could also ensure active participation of policy-makers and representatives of R&D institutions. METI through its network could attract researchers in the field and also representatives from international funding agencies like UNDP, IDRC, etc.

The present volume is a direct offshoot of the above mentioned workshop. The issues which have been debated and highlighted in the workshop provided the background for the studies incorporated in the volume. The volume has been planned to put together some of the research insights for Indian situation in the context of globalization along with the experiences of countries in Latin America which face more or less a similar dilemma and South Korea which has successfully managed to make a mark in the world economy by carefully strategizing its Science and Technology (S&T) and economic options.

Management of technological capability building requires a long-term strategy. An important part of such a strategy is to develop close linkages and partnership between R&D and industries. In a typical case of India, as argued by G.D. Sandhya, technological priorities always remained subservient to short-term policies for economic crisis management. Thus, in a situation of balance of payment crisis, import substitution becomes a policy priority. The R&D system, therefore, is geared to import substitution technology. How does the technology system respond once the balance of payment crisis is resolved and the policy of import substitution is underplayed? More than anything else, it creates a situation of confusion for the technology system which can never restructure or reorganize itself to be tuned to the ever-changing economic priorities.

Linkages between industries and R&D system have to be long-term in nature and free of the vagaries of short-term economic and political fluctuations. Alberto Araoz draws from a study on Latin American countries for revitalization of Industrial Technology Research Institutions from the perspective of close and durable interaction with industries.

The South Korean experience in this regard has been different. Through government initiatives efforts have been made to promote long-term cooperative research between industry and/university/ R&D institutions. Chang Jae Lee and Deok Soon Yim discuss the success factors behind such cooperative efforts.

In the same tune, N. Mrinalini examines the experiences of countries like Taiwan and Japan where government initiatives could bring together both the user and generators of technology to contribute to technological capabilities of targeted domestic industries. The study brings out the possible role of government in a situation of globalization as the role of initiator and facilitator rather than an interventionist.

What is the difference between Kuwait and South Korea? South Korea has much less per-capita income than Kuwait. Can we, therefore, conclude that South Korea's technological capability is less than that of Kuwait? Capability in this regard has to be seen in terms of the respective source of their GDPs. Kuwait is largely dependent on its natural resource reserve, whereas South Korea is deprived of any significant natural resources. South Korea's main strength lies in the level of skill of its manpower which is much stronger than that of Kuwait. South Korea's economic achievement, therefore, is sustainable on a long-term perspective, while that of Kuwait is not. The point we are trying to make is that the real wealth of a nation in a globalized economy is its manpower and its level of skill. Technological capability building, therefore, needs a long-term strategy for the skill development of a country's population in targeted sectors and fields. B.R. Virmani's study highlights how technology transfer is dependent on the quality of human resources available in a country. The issue gets paramount importance since success in technological capability building depends not only on developing technologies but also more on learning technology through technology transfer.

How did India perform in this regard? S. Visalakshi's study on human resources development in biotechnology suggests that what precisely Indian policy in biotechnology lacks is a careful synchronization between industry's growth and matching manpower requirements. This is conspicuous by the presence of mismatch between type of manpower demand and available manpower in this new technological field.

The case of paper industry is no different. Indian paper industry has remained technologically laggard since the beginning. Is it impossible to target this sector, develop a long-term strategy, mobilize all technological and R&D resources to uplift its standard at the global level? N.J. Rao's study is affirmative about the potentiality, only if certain key issues are taken care of.

Three case studies by Shekhar Chaudhury, Santanu Roy and Tabassum Jamal focus on the issue of R&D industry partnership and collaboration. Chaudhury and Roy studied cases of a few projects of R&D laboratories. Their studies suggest that failure or success of an R&D project can clearly be attributed to the intensity, extensity and nature of interactions with the prospective users of the research results. Wherever such interactions have been in-built in the project the research results could easily lead to end use. Jamal's study highlights a similar issue for the paper industry. The successful technological ventures, in the cases studied, have always been the result of collaboration among firms and between firms and research institutions.

Finance holds the key to strategic ventures. Even in the literature on economics of technological innovations finance is the most neglected aspect. A management view of technological capability building would give centrestage to the financial aspect of the problem. In developed countries the finance sector developed almost hand in hand with the new technological ventures. As Mrinalini's study has shown, much of the success of Taiwan has been due to institutionalizing finance with R&D and applications. Venture capital type of arrangements holds a significant promise for financial arrangements for promoting new technological ventures. The Indian performance in this field, however, is not remarkable. Sunil Mani has elaborated the constraints and prospects of venture capital arrangements in India in the light of the experiences of the developed countries.

Two studies in this volume discuss the technology management and R&D related issues of the corporate sector — the user side of technology. F.M. Machado's study is based on United Nations Industrial Development Organization (UNIDO) project aimed at Technology Management and Industrial Informatics for the Industrial Modernization of Costa Rica. For this purpose a centre (Centre for Technology Management and Industrial Informatics) was created to assist Small and Medium Enterprises (SME) with training, information and consulting services in technology management. The study details the barriers encountered for the introduction and adequate application of technology management concepts and methodologies, and the specific solutions arrived at.

S.D. Moitra's study reviews the structural strategic, and tactical factors responsible for successful R&D in terms of commercialization. The focus is drawn on in-house R&D of the corporate sector. The success factors, however, are the same, Moitra argues for the non-corporate R&D as well, because most of it depends upon the right choice of a R&D project based on market feedback along with proper placing of manpower and project management.

The last contribution in the volume by P. Banerjee deals with the contemporary problem of R&D in advanced engineering and information-intensive R&D. Not only industrial culture but the very nature of R&D has also undergone considerable change. It is the phenomenal developments in information technology that have changed and are continuously changing the industrial and R&D practices. The main feature of this change is simultaneity in production and R&D. These activities can no longer be compartmentalized. The cultural change it has brought about is the essentiality of cooperation and partnership among R&D and the production system. Banerjee analyses the case of aeronautical research to focus on this emerging issue.

1

Institutionalization of S&T Planning Process in CSIR: The Policy of Import Substitution and Its Strategization

G.D. Sandhya

In today's context of globalization, when research organizations are being restructured/reorganized, institutional frameworks are being evolved for technology development, and industrial and technology policies are being reformulated, it has become necessary to try and take stock of how these research organizations have performed during earlier policy regimes. The success of industrial development in East Asian countries can be attributed to institutionalization of the whole process of cooperation of R&D institutions, industry and proper technology and industrial policy instruments, through which these countries encouraged mastering of imported technology in sectors of their comparative advantages. In India, too, we have successful cases such as space, atomic energy, agriculture, defence, etc., the success of which can be attributed to the clarity of their mandate and sustenance of functions over a long period of time due to a strategic vision and also due to their integration vertically. The performance of the Indian industry, however, leaves much to be desired despite the fact that India has an impressive chain of government funded industrial research laboratories doing industrial R&D. Now when globalization has necessitated a need for rethinking, let us look at how policies are strategized. One can visualise the whole process of institutionalization of any policy command or directive from two angles. While the one refers to the formulation of policies with a holistic approach and a perspective, the other may refer to policy implementation. This study deals with the former and tries to deal with the implications of adequacies or inadequacies of policy formulation. As the process of Science and Technology (S&T) planning is operationalized at the national, organizational, and institutional level, it is necessary to see what kind of goals and targets are set at the national level. In this context it would be worthwhile to look at how research institutions operationalize the goals that are set out by the State.

Import substitution can comprise both fiscal component, at one end of the spectrum, and capability building through a sustained learning process at the other. The proponents of this strategy have believed in the strategy to encourage indigenous efforts and talent, entrepreneurship, skill, self-reliance, domestic technological development and suitable range of products [Ballasa 1970]. The adoption of this strategy as a mode of development for the developing countries has been supported by many [Chenery,1960; Myrdal,1966]. The critics have believed the policy to be inappropriate because of its inability to create cost-competitive economy, move into export markets and to destroy the incentive necessary for improving the quality of output [Hershman,1968; Prebish,1967]. The import substitution policy was consciously adopted in India because of the deteriorating balance of payments and a deep faith of her planners in self-reliance. This study examines the policy institutionalization process by taking the case of the Council of Scientific and Industrial Research (CSIR), which promotes the development of S&T for industrial development and both is funded and administered by the government. The onus of guiding and planning the research of CSIR lies with the State which issues directives from time to time. This study begins with a review of the S&T planning process to glance over the kind of national priorities that have been set by the Planning Commission over the years. An attempt is made in the following sections to look into the management of one of such policy directives, i.e., the import substitution directive and its implementation in the CSIR. The issue gains importance from a point : what lessons do we draw through this ex-post evaluation in the context of the present restructurization of CSIR? The success of any restructurization exercise hinges very much on its strategization. Import substitution as pursued in many other countries such as Japan and South Korea is surely a positive case in point. The strategization of import substitution there involved government industry and other agencies for its successful culmination. The basic difference between them and India has been with regard to the strategization of the policy formulation which necessitated the need of research by these countries on what was imported by the industry. On the other hand, this process of strategization in India allowed industry and R&D institutions to work in isolation. This study provides evidence on the process of import substitution policy formulation for R&D in the CSIR.

Policy Signals for the Planning of S&T

The Planning Commission since its inception on 15th March, 1950 has been preparing and coordinating five-yearly national developmental programmes. About CSIR as we are aware, a noteworthy point is that though autonomous, it is supposed to plan its activities in the framework of the priorities evolved by the Planning Commission. Now we will very briefly look at what directions have been given for the industrial research to CSIR by the Planning Commission at the national level, since its inception.

The First Five Year Plan assigned a mix of roles to the CSIR from institution building, resource surveys, work on basic and applied research, new and known processes and products and cost reduction to its role towards small and medium scale industries, because research facilities were lacking in these industries.

No direction was articulated in terms of a clear and coherent strategy. While in the First Five Year Plan, attention was chiefly devoted to the building up of national laboratories and other research institutions, the primary object in the Second Five Year Plan was to develop the existing facilities and to bring the work of the scientists in the national laboratories and of research workers in universities and other centres to bear, as closely as possible, upon important problems in different fields of national development.

The principal aims of the Third Plan were to achieve self-sufficiency in food and expand basic industries. The role envisaged for scientific research was to cover the distance (the gap between the advanced and less advanced countries) by putting in the utmost effort in the development of scientific and technological research and in the application of science in the furtherance of the development programme. The multiplicity of goals got carried over from the first two Plans to the Third Plan. The role of CSIR was, however, expanded during this period due to defence crisis and deteriorating balance of payments. The CSIR was asked to reorient its research to help in defence requirements. The foreign exchange crisis tilted the balance in favour of import substitution research during mid 60s. The National Council for Science and Technology (NCST) reaffirmed the stress on programmes of major socio-economic impact. The overall emphasis was placed on minimizing imports, promoting exports and enhancing productivity. The strategy was to pursue and complete projects of short duration, enabling quick results.

The priorities in the late 70's were rural development, emploment generation, poverty alleviation, etc. A role was visualized for the CSIR for the provision of minimum social service for those below poverty line in the area of water supply, health, housing, etc. The emphasis shifted to basic research, technology missions and frontier areas. The CSIR was supposed to be working for the missions that involved large S&T components. The developments in the 90s at the political and economic front require that CSIR should now work in a globally competitive environment and prepare itself for the transition from centrally planned economy to market-driven economy. What has been noted is that goal setting, as it appeared, looked like a statement of good intentions but institutionalization of such goals through clear-cut strategies is yet to be seen as we are going to see in this study. It is clear that many initiatives were taken quite frequently to reorganize and restructure R&D institutions. The manifestations of these shifts at the CSIR level are often unclear. It is clear that there is difficulty in relating these plans in a coherent and organic way. There have been frequent changes in the plans lacking clarity. The same has been manifested at the CSIR level [Sandhya, 1990]. There are difficulties for institutions for deriving S&T priorities from national level plans, if they are not strategized properly. This is coupled with the fact that we have not been successful in creating a demand for indigenous technology, in the absence of which, even factors of linkages between demand and supply have failed to complement R&D and production.

The next section illustrates the process of institutionalization of import substitution research in CSIR, its management, i.e., under what circumstances import substitution directive was issued to the Council, how the Council perceived, interpreted and passed it on to the constituent laboratories.

Evolution of Import Substitution in CSIR

In what succeeds, the genesis of import substitution policy in the CSIR is traced with a brief deliberation of the circumstances under which the policy was drafted, directed to CSIR and therefore, to the research institution under consideration, i.e., the National Chemical Laboratory (NCL). A clear-cut emphasis on import substitution research followed as a consequence of the foreign exchange crisis of 1964-65. A country-wide appeal was made towards the end of 1965 towards import substitution and to bring together the country's research facilities and the industry's know-how requirements [BSIR, 1965; G.B., 1965; D.C., 1965]. It was in 1964, when the Vice-President of CSIR, Shri M.C. Chagla, the then Minister of Education, while presiding over the Directors' Conference called for top priority to be given to those research activities which were aimed at finding ways and means of conserving foreign exchange potential for the purpose of making the country self-reliant to the maximum possible extent [D.C., 1964]. He advised the directors to reorient the research programmes and show how efforts of research would help to conserve the foreign exchange. A working group was formed on the planning and organization of research under the Chairmanship of Dr Amarjit Singh. The working group noted with concern the need for conserving foreign exchange. The governing body after ten months concerned itself with listing the research projects and pilot plants in the laboratories under (i) defence, (ii) import substitution/elimination, (iii) industrial technology, (iv) food and agricultural products and (v) objective basic research. A statement giving the total number of projects and pilot plants in the National Laboratories and their orientation under these heads was made. Even the Fourth Plan proposals of CSIR were discussed at length within the same framework. It was decided to make a critical study of ways of and means of bringing down the foreign exchange content of the 4th Plan to Rs. 41 crores. It is interesting to note the importance of export promotion almost on a similar note as of import substitution, for instance, Vice- President CSIR considered export promotion as being equally important from the point of earning foreign exchange [BSIR, 1965].

A gettogether of research institutions and industry was organized by CSIR and industry in December, 1965, in pursuance of the directive the of governing body of the CSIR [CSIR, 1966]. The gettogether showed its concern to the problems of (1) development of substitutes for imported raw materials, (2) indigenous manufacture of essential components hitherto imported, (3) development of indigenous know-how at a sufficiently advanced level as to be commercially utilizable in respect of those articles on which repeated foreign collaborations had been sought, and (4) measures through which the programmes of industrial research could be oriented to the needs and requirements of industry. The gettogether strived for the reorientation of research programmes of the laboratories to have a direct relation to and bearing on the Industrial and economic plans of the country. The two gettogethers, one between CSIR and industry [CSIR, 1966], the other one by ICMA (Indian Chemical Manufacturers, Association) [ICMA, 1968], discussed problems of import substitution. The first gettogether identified 237 projects as nationally important. These two get togethers identified projects which could be launched in the CSIR laboratories. This was an attempt to reorient the research programmes of the laboratories to have a direct relation with the industrial and economic plans of the country. Roughly 400 papers were discussed during this get-together amongst scientific and technical personnel from industrial firms and government laboratories. At this time nearly 400 projects dealing with substitution of raw materials/imported components or alternative technologies to imported know-how were in progress in the laboratories. It was felt that the effectiveness of such a sizeable effort could be further improved by adding selected projects dealing with substitution of imports of high value of strategic nature to the existing lot.

What is of interest here is that the papers presented in the gettogethers defined broadly national tasks in terms of physical gaps of imported raw materials, plant and equipment, machinery and strategic items.

The Nature of Import Substitution

It is pertinent to note the developments in the whole process of the understanding of import substitution. The shaping up of the strategy of import substitution was based on the considerations that are delineated in the following paragraphs.

Regarding the nature of import substitution it is relevant to note the views of the key persons involved in the shaping up of the strategy of import substitution. In the Research and Industry get-together, the Chairman, Vice-President, CSIR, gave the following suggestions in his address:

> "The first and foremost suggestion I would like to make is precisely the same as whatever the Prime Minister has been saying from time to time emphasizing the importance of scientific research which we have to adopt as a slogan that is "self-reliance". It is not something you say on the platform but it has to be translated into concrete terms as far as industry is concerned. Self-reliance can only come about if you make up your mind to do without imports or do with as little imports as possible. It means to try and find out whether you can substitute for the things which we are importing."

Reflections on the understanding of the whole process of

import substitution in this framework can be easily seen in the criteria used for the selection of R&D projects. Here it is pertinent to note that the Ministry of Commerce had formed a committee under the Chairmanship of Prof.V.K.R.V. Rao, on import rationalization and import substitution towards the middle of 1965. While recognizing the progress in the form of appreciable work being done in the last 15 years with regard to import substitution in the country, he commented that the problems of import substitution had not been tackled in a comprehensive and scientific manner and that there was a large scope for import rationalization and substitution in various directions.

Efforts were made to concentrate on its working out strategies for time-targeted and nationally important projects, in the next directors' conference [D.C.,1965]. The conference discussed the mechanism for implementation of those 237 projects which were identified as being nationally important. This was set out in detail in a position paper prepared by Dr T.S. Subramaniam and discussed amongst the directors. The following observations were made to serve as guidelines for the formulation of research programmes in the laboratories by the President, CSIR, in the governing body meeting:

1. Emphasis was to be laid on specific projects which had a direct bearing on the practical needs of the country.
2. Emphasis was also to be laid on projects which could be completed quickly within the limited resources available.
3. It was to be clearly thought out as to what was of absolute priority and what could be put off [G.B., 1966].

The President, CSIR also emphasized on coordinating research well with the Five Year Plans of the country. The action taken on the recommendations of Research and Industry gettogether of 1965 was the subject under discussion at the next governing body meeting. Action was taken in two parts mainly:

(1) action in regard to R&D policies and technical recom-

mendations in respect of processes and products which do not require R&D; and

(2) projects requiring high priority research and development work [G.B., 1967].

Action was initiated on every recommendation and correspondence initiated. The recommendations on which action was processed related to the setting up of technical research committees which were under consideration of the developmental councils, rationalization of import duty on raw materials, intermediates and finished products, etc. The meeting of chairmen and conveners of the 15 working groups held in February 1966 consolidated the priority research projects into 237 items. On scrutiny, the number came to 168 for which proposals were sent by the laboratories, industries and universities. The indications were that the emergence of import substitution research was more of an instrument to conserve the foreign exchange in a very conservative sense. The next section provides some evidence on how the import substitution got implemented in an institution of CSIR.

Import Substitution in NCL

The laboratory was established in 1950 with the objectives of creating research facilities in the chemical sciences and chemical technology as were envisaged by the planning committee[1]. An attempt is made in the following paragraphs to see what sort of import substitution research was carried out in National Chemical Laboratory during this period, what were the factors determining the choice of research programmes or what justifications were given for undertaking research programmes after the directive of import sub-

1. The mandate of the laboratory was to do fundamental research for its own sake and to extend the frontiers of knowledge in directions likely to prove useful for the industrial progress of the country, applied research of high order, provide a link among university, the State and other scientific institutions, etc. [CSIR,1945]. The first fifteen years (1950-65) were spent in building up a high order of scientific research capability in chemistry and chemical engineering by concentrating on basic research.

stitution was issued. A quantitative assessment of the R&D projects undertaken during this period revealed that the concept of import substitution manifested itself in the form of product-oriented schemes. The agenda of research project prior to this period had mostly discipline-oriented research programmes with emphasis on basic studies, out of which technologies for, say products, got generated. It is only during this period, one finds, most of the projects were selected around products. To cite an example, the area of, say, catalyst development and their evaluation for petroleum refining can include studies on physico-chemical properties, compositions, fundamentals of chemical action shown by various transition metals, etc., on the one hand and on the other, it can include projects on the development of technology for the catalysts already known. The latter constitutes the product-oriented schemes. As per this classification, product-oriented schemes in the laboratory included (1) products which were imported in the country, there being no indigenous production, (2) products which were produced in the country but whose production was insufficient. To fulfil the leftover demand the R&D work was initiated in the laboratory.

A total of 410 product-oriented schemes were undertaken in the NCL during 1965-78, of which 203 were of import substitution in nature constituting roughly 50 per cent. An examination of research proposals revealed a gradual increase in the number of product-oriented as well as import substitution schemes from 1965 to around 1970 (Table.1). There was a gradual decline in the number of product-oriented projects from 107 in 1974-75 to 76 in 1977-78.The decline in the number of import substitution project was from 44 to 25 in the same period. There was a sharp decrease in the number of product-oriented schemes from 1978. The number of product-oriented schemes came down to 36 in 1978-79 [NCL,1979] and 25 in 1980-81 [NCL,1981]. An appreciable increase in the number of import substitution projects from 1965 onwards indicates that the import substitution directive was definitely instrumental in converting research programmes to product-oriented research projects, which were in turn of import substitution nature. The evidence suggested favour for those projects which led to substitution of imports and those for which no capacities existed in the country. Even those products for which some productive capacity was available with industry were included in the research programmes of the laboratory, e.g., the project on chlorination-hydrochlorination of ethylene were justified on the grounds of supplementing the existing capacities [NCL, 1967].

Similarly, the whole programme on organic chemicals was undertaken in the laboratory as a massive import substitution programme. The basis of selecting this programme was that a number of organic compositions available under trade names found special applications in different industries. They were imported and their analysis to isolate and identify components was undertaken with a view to seeing the feasibility of manufacture of similar compositions from indigenous raw materials. By 1968-69 this group diversified from analysis to technical preparation of organic intermediates. Work began on developing know-how for the manufacture of a variety of organic chemicals produced by batch processes and which were imported in substantial quantities. They were listed in the import statistics under NES (Not Specified Elsewhere) category. Processes were to be selected on the basis of imports and demand pattern in the country and interest shown by the industry for commercial exploitation [NCL, 1969]. The advisory committee recommended the programme to be undertaken on a priority basis. The tentative list included polygard, methyl phosphate, xanthates, para-nitrophenol, tertiary-butylcatechol, fungicides, weedicides, insecticides, etc. Another notable thing is that equipment to cover processes of sulphonation, nitration, reduction, alkali fusion and halogenation as batch process were ordered under capital grants sanctioned under this project.

Table 1
Product-oriented and Import Substitution Schemes Undertaken in NCL from 1965-78

Year	*Number of product oriented schemes*	*Number of import substitution schemes*	*% of import substitution schemes*
1965-66	50	8	16
1966-67	93	36	39
1967-68	96	39	41
1968-69	117	51	44
1969-70	103	54	52
1970-71	96	46	48
1971-72	118	54	46
1972-73	96	43	45
1973-74	98	38	39
1974-75	107	44	41
1975-76	98	37	38
1976-77	83	32	39
1977-78	76	25	33

Source: National Chemical Laboratory.

A review of research proposals revealed that a majority of research projects were justified on the basis of import substitution of either the technology or the products based on that technology. As we looked into the minutes of GB and Board of Scientific and Industrial Research (BSIR) meetings along with the minutes of the Director's conferences, we did not find any evidence on nuances of the adoption of import substitution and its implications on the capability building in CSIR. The discussions revolved around fiscal issues rather than the issue of capability building. As the nuances of import substitution directive were derived from national policy pronouncements, the institutional interpretations reflected the adoption of strategy without incorporating the element of capability building on a long and sustaining basis[2]. The adop-

2. It may also be noted that the work and functioning of CSIR has been reviewed from time to time by various review committees. The initial criticisms of the NCL were with regard to its little impact on industry by the Third Review Committee largely due to its continued emphasis on basic research. The laboratory was asked to alter the balance between fundamental and applied research from 50:50 to 20:80 respectively. In fact, this review was followed by the period of exchange crisis. It seems probable that these two necessitated the need for legitimizing the research projects on account of their applied nature.

tion of import substitution could be related more to its use as an instrument of conserving the foreign exchange rather than an instrument of achieving self-reliance.

Data analysis is done to show the type of opportunities for long-term R&D capabilities that got missed out or strengthened due to the adoption of import substitution. The R&D projects of NCL are analyzed for the entire period commencing with the definite assertion by the government on import substitution in the 60s to innovative technologies in the 80s. Now we will look at what were the implications of doing this kind of research in the government-funded research institutions which are, in principle, expected to perform long-term targeted research. This essentially means that any attempts in orienting the research should have necessarily incorporated the very basic element of capability if it were to be treated successful and that is how the assessment is done of the success or failure of import substitution strategy as well.

Implications of the Adoption of Import Substitution Research

Now let us understand the implications of import substitution research performed by these institutions on a long-term capability building. Import substitution can be direct and functional in nature. The former refers to defensive import substitution, whereas the latter has both economic and technical aspects and refers to potential import substitution. The success of the latter depends to a large extent on technology factors and thus R&D in the application of new technology which in turn contributes to strengthening of the R&D capability of the laboratory. A potential R&D programme of a laboratory enriches both the laboratory and the user. This is the best contribution of an industrial research laboratory. What is important is the fact, that even if an R&D programme does not enrich the user, it must enrich the laboratory, e.g., if a laboratory specializes in a reaction mechanism such as fixed or fluid-bed operation, the capability of specializing in the reaction mechanism can be used in other production processes. It is here one comes across technological or functional import substitution, which means alternative techniques of production, substitution of one product by another of different identity but performing the same function. This is the most important aspect of import substitution because of both technical and economic parameters. The adoption for either kind of substitution can have implications for capability building of the laboratory. The implications of the adoption of import substitution research are manifold. The projects can be classified into the following two categories:

The first category refers to technological import substitution. As already discussed earlier, the strength of a national laboratory lies in building up the expertise, based on which even product know-how can be generated. The scope of the term technological import substitution is broadened in the context of R&D; for instance, there are two ways of selecting an import substitution project— one is that imitative research is undertaken without basic/applied basic research; the other is that import substitution project comes out from a well-built area of basic/applied basic research work. The latter can provide a good synthesis of basic, applied and developmental work, as we shall see in the following paragraphs.

The basic work on silicon chemistry was pursued in the NCL for almost two decades, before the projects like polycrystalline silicon and chlorosilane were undertaken in the laboratory. These were the projects involving high innovative component. The work on titanium chemistry, dye chemistry, surface chemistry, and polymer chemistry can be mentioned here to highlight the case. At present NCL has a strong base for R&D in catalyst development and evaluation for petroleum refining and organic chemical industries. Two and half decades ago the objectives of this programme were of basic nature only which ranged from studying physico-chemical properties of catalysts, their compositions, fundamentals of chemical action shown by different transition metals, preparation of the catalysts, to studies on hydrogenation kinetics to evaluate their activity and mechanisms of reactions, etc. The programmes during late 60s up to around mid 70s received hardly any attention in terms of priorities, but NCL did not work on many product-oriented import substitution projects in this area. The way basic research for capability building in selected areas can lead to advancement of knowledge for further industrial growth can be best exemplified by this case. The programme right from the beginning was tilted towards basic research. The process development work in the laboratory involved usage of molecular sieve catalysts developed in the physical chemistry group. The other studies related to stability of zeolite catalysts in systems, viz., alkylation of benzene and disproportionation of toluene and development of catalysts in butenediol production and for aniline, etc.

The contributions of these basic research programmes to applied research are as follows: (i) development of novel non-noble metal catalysts (zeolites) manufactured by ACC, commissioned in the xylene isomerization plant of Indian Petrochemicals Limited (IPCL). These catalysts are superior to conventional catalysts. Now NCL is in a position to offer complete technology package for isomerization of xylenes in collaboration with IPCL and Engineers India Ltd (EIL); (ii) another catalyst Encilite 2 is for the production of ethylbenzene from ethanol and benzene in a single step. The catalyst, to be manufactured by ACC, will be utilized by Hindustan Polymers in the 14000 TPA plant having a turnover worth US $ 13 million; (iii) world class catalyst Encilite 3 for toluene disproportionation will be manufactured by ACC. EIL- designed plants for toluene disproportionation are likely to be commercialized by IPCL and ACC; (iv) another catalyst, Encilite 3, has been tested by an American catalyst manufacturer in a Mexican refinery which is negotiating for the purchase of this catalyst worth US $ 8 million.

Similarly the programme on polymers began as a consolidated programme of product and process development including reaction engineering and rheology. Basic studies constituted an integral part of the total R&D programme which ranged from physico-chemical studies, studies on thermodynamic properties, polymer characterization, viscoplastic properties, to studies on polyamide and dimer acid, etc. The programme has resulted in the development of a number of state-of-the-art technologies. To name a few,

(i) The technological monopoly of sulphochlorinated polyethylene (SCPE) is held by a multinational company. Based on the data generated by a pilot plant at NCL, Shriram Rayons is setting up a 150 TPA plant as a precursor to 3000 TPA commercial plant of the chemical

(ii) Indian Oil Corporation Ltd. has set up a 200 TPA semicommercial plant for manufacture of Jalshakti and obtained government clearance to set up a commercial plant of 5000 TPA having a production value of about 33 million US $.

(iii) IPCL has set up an integrated 10,000 TPA plant, with an investment of US $16 million for the manufacture of acrylates from acrylonitrile. This joint NCL, IPCL, and EIL technology enables the manufacture of all types of acrylates with only two sets of equipment.

The case of vitamin B_6 may also be cited here. India's efforts were aimed at replicating the process in 60s but by the time the laboratory stabilized the process, a better and more efficient process had been commercialized by a multinational company. The restart of work in the laboratory in the 70s was again to stabilize the known methods of vitamin B_6 production and determine process parameters for a manufacturing process. The project again met with failure. Efforts were made in the 80s to develop an economically viable alternative process for the production of vitamin B_6. For the alternative route, the work on vitamin B_6 began from the efforts of the scientists to understand the chemistry part of the process. The plant was commissioned in 1987 successfully.

Many other important projects were developed this way on a solid foundation of basic research. These led to the development of potential technologies and also substituted imports. To name a few, acrylates, endosulfan, butyl rubber, dimer acid, polycrystalline silicon, and anisidines, ethylenediamine., etc., were developed subsequently

The second category refers to defensive import substitution. In this category we include the projects which do not add to the capability building of the laboratory. This we call single product projects, which do not help in strengthening the capability of the laboratory to which the laboratory is committed. The best place for undertaking R&D for these projects could have been the in-house R&D units of the industry, e.g., projects on aluminium silicate as rubber filler, maleic anhydride, phthalic anhydride, aluminium hydroxide gel, thioglycolic acid, triphenyl phosphite, nonyl phenol, can-sealing compositions, etc. These were undertaken in the laboratory on the basis of their import statistics. These projects did not add to the stream of expertise building in the laboratory. Similarly, other projects that can be put in this category are the projects which were not relevant to the mandate of the laboratory, e.g. the whole area of prototype developments of various chemical instruments. A majority of the applied research projects were taken up during import substitution period, e.g. ultraviolet spectrophotometer, infrared spectrophotometer, chromatograph, etc. Here sustained development and modernization of the prototype was not the goal of the laboratory, which meant that after the development of the prototype in the laboratory, the work came to an end.

Summary and Conclusion

The endeavour clearly reflects that there existed no effective system of strategizing the policy commands/signals. Import substitution policy was poorly articulated, and badly managed, which resulted in mixed benefits with setbacks. Import substitution came in as a directive but the concept never got defined meaningfully. The conceptual formulation of import substitution did not become the subject matter of discussions at any of the planning levels. Import substitution was equated to product synthesis. As a result, a large number of product-oriented schemes were picked up in the laboratory. This resulted in an undue emphasis being placed on the projects, many of which could have been carried out in the industrial R&D units easily.

In India, import substitution as given to the research institutions provided a culture, where it was always a victim of "lagged time" in respect of technology opportunities and application. In other words, as a corollary it implied that a laboratory shall wait until a discovery is made, an invention emerged out of it, the invention is commercialized, the product marketed, and at this stage the laboratory was told to catch on to a substitute or domestic market. Also the Indian situation is different, where research was not done on what was imported, like Japan but on what was not imported. Thus, the role visualization of research institute was that of an institute which will develop either the technology developed elsewhere or substitute the product imports. The basic argument of the study is that the strength of such specialized research institutes does not lie in serving short-sighted gains, rather it lies in building up a long-term scientific capability.

It is possible to discern the implications of the import substitution research in view of the above classification. The involvement of the laboratory for defence import substitution projects had implications for the capability building of a laboratory. The inclusion of single product projects did not lead to a continuum of activities for the planned development of an area. Here the effective optimal utilization of resources, scientific manpower and equipment could have been accomplished in a number of cases by sharing the research effort for process parameter generations for specific products between the NCL, exercising the role of a research counsellor and industry developing the process.

The most inherent flaw of the policy was that the components of import substitution were not defined. The stress was not laid on technological import substitution, which in combination with applied time-targeted projects could have consolidated the existing capability of the laboratory. The development of a range of catalysts in the 80s is an example of the generation of S&T strength in the area. This contributed to another dimension of import substitution in the sense that these catalysts have found an export market. There should be a continuous effort to evolve strategies for S&T planning and make structures to implement the strategy.

The process of S&T planning underwent many changes. There are clear reflections of linking R&D planning with unconnected priority tasks and short-term political changes. The broad conclusions are that in India the policy of import substitution just failed to provide a dynamic perspective to the research institutions, unlike other successful import substituting economies such as those of South Korea and Japan. The strategization of import substitution there involved government industry and R&D institutions with a clear perspective. The basic difference between them and India has been with regard to the strategization of the policy formulation which necessitated the need of research by these countries on what was imported by the industry. On the other hand, this process of strategization in India made industry and R&D institutions work in isolation.

REFERENCES

Balassa, Bela (1970). Growth strategies in semi-industrial countries. *Quart. J. Econ.*, 84(1), pp. 24-27.

Chenery,H.B.(1960). Patterns of Industrial Growth, *Amer. Econ. Rev.*, p. 640.

Myrdal,G. (1966). *Asian Drama, An enquiry into poverty of nations* (Penguin Press, London, 1966).

CSIR (1945).Report of the Industrial Research Planning Committee, New Delhi, CSIR.

CSIR (1966). The first Gettogether of Research and Industry, 20-21 Dec., 1965, CSIR, New Delhi.

BSIR (1965). Proceedings of the 55th BSIR meeting 6 Oct.1965.

D.C. (1964) Proceedings of the 14th Directors., conference: 4-5-7 Dec., 1964.

D.C. (1965) Proceedings of the 15th Directors conference: 19 and 22 Dec. 1965.

G.B. (1965) Proceedings of the 50th Governing Body Meeting: 7th Oct., 1965.

G.B. (1966) Proceedings of the 52nd G.B. meeting: 19th Dec., 1966.

G.B. (1967) Proceedings of the G.B. meeting: 25 Nov 1967.

ICMA (1968). Proceedings of the seminar on the role of industry and the role of R&D in import substitution organized by ICMA., 10- 29-30 August, Bombay.

NCL (1960-1988). National Chemical Laboratory, Annual Reports, and Research Programme Documents, 1960-88, Pune (India).

Prebish,Raul.D.(1967). Towards a new trade policy for development, UN Hershman, A.O.(1968) The Political Economy of Import Substituting industrialization in Latin America, The *Quart. J. Econ.*, 82(1), pp. 13-15..

Sandhya, G.D, Jain, Ashok, Mathur, Purnima (1990).'S&T Planning Policy Direction and CSIR, *Econ. and Pol. Wkly*, No.51, 25(51) pp. 2800-2805.

2

Industrial Technology Research Institutes in Latin America*

Alberto Araoz

Industrial Technology Research Institutes in a Changing Context

There is concern in many developing countries about the relevance and performance of their scientific and technological institutions, in particular those devoted to industry — the Industrial Technology Research Institutes (ITRIs), or Research and Technology Organizations (RTOs) in the nomenclature adopted by World Association of Industrial and Technological Research Organization (WAITRO).

ITRIs are often seen to have only limited impact, and there is a feeling that the financial and human resources invested in them should be used as efficiently as possible in support of national goals. This will often need a restructuring and reorientation of such institutes.

At the root of this general issue are the major challenges now being faced by developing countries, among them: accelerated technical change; the increased importance of trade in a highly interdependent world with globalized production systems; the liberalization of the economy and the abandonment of import substitution; the shrinking role of the State and the privatization of a good part of the public sector; a change in the attitude towards foreign investments, which is increasingly welcome as a source of capital

---

* The subject is dealt with more extensively in my report "The Revitalization of Industrial Technology Research Institutes in Developing Countries", submitted to UNIDO in 1994.

technology, market access and management know-how; and increasing concern about the environment, and others that distinguish the world of the '90 from that of previous decades.

As a consequence, ITRIs in the developing countries are now facing several interrelated issues shown as follows:

(a) *Relative failure to fulfil their original objective of assisting the technological development of industry:* Many ITRIs have tended to put emphasis on topics of their own choice in their R&D activities, and have not been successful in the commercial application of research results. By not focusing closely on the requirements of industry, they run the risk of becoming largely irrelevant to the development needs of industrial enterprises. This risk is reinforced by their lag to master new technological fields and management techniques that are becoming increasingly important for industrial enterprises.

(b) *Different requirements stemming from the changed milieu:* The changing technical, economic and policy environment of the nineties makes life difficult for the traditional ITRIs, which were attuned to import substitution industrialization. They now have to assist industry to become much more competitive in order to participate in markets that have opened up and are no longer strongly protected, and to step into the international market. The new milieu would seem to require a different type of ITRI, more flexible and better able to help industry respond to this new set of challenges.

(c) *Dealing with new types of problems:* ITRIs are increasingly under pressure to address problems that are new to them—advanced technologies, productivity enhancement, design, quality, management of industrial enterprises, environmental protection, etc.

(d) *Helping the transition from techno-nationalism to techno-globalism* : Developing countries and the enterprises operating in them should increasingly consider the need to search for alliances and cooperation beyond national borders, if they are to have access to the technologies of the most dynamic sectors, and to avoid an ever deeper marginalization. These countries have not yet entered in a significant way the world markets of products and services that employ advanced technologies. The acquisition of foreign technology, as it has taken place traditionally in the past, seldom allows local industrial firms to obtain access to the most up-to-date technologies. On the other hand, it is not easy to carry out on a strictly national level the efforts required to develop new products and processes that may open up new exports in the most dynamic sectors of the world market. In most industrial sectors, the efforts of enterprises do not seem to be enough to close the technology gap, and develop internationally competitive products, even when they cooperate among themselves and with the scientific and technological institutions within the country. The activities of the "national innovation system" are far from sufficient in the technologically advanced areas. Techno-nationalism should give way to techno-globalism. Enterprises wanting to operate efficiently in those areas should join in the globalization of technological efforts, through strategic alliances and collaborations with world technology leaders. This is being done in several countries of recent industrialization in the Asia-Pacific region which have been able to build up a well-trained workforce and a scientific and technological capability of such a level that it can cooperate on equal terms with the leaders in certain areas.

In this way, enterprises can participate in the development of new products and processes and in collaborative schemes of production of high technology goods. Other developing countries should attempt to follow a similar path to avoid marginalization in future, and ITRIs in those countries should help the transition from techno-nationalism to techno-globalism.

(e) *Need to update capabilities and develop new skills:* In order to cope with the new technological realities and the resulting requirements of industry, ITRIs find that they need to update their scientific and technical manpower, their procedures, their management methods, and their installations and equipment. This means, very importantly, the introduction of new knowledge and skills through training their human resources, selectively hiring new personnel, using capable and experienced outside consultants in their projects, improving their information resources, going into collaborative research projects, establishing strategic alliances with other institutions at home and abroad, and so on.

(*f*) *Reduction of State support:* Most ITRIs face worsening conditions as their State funding has shrunk, often on account of stringent fiscal policies that reflect the declining role of the State, while the demand from industry has seldom grown enough to compensate for the loss of revenue. This situation has led to poor salaries, lack of adequate library facilities, instruments and equipment, and in some cases a succession of crises which have discouraged the scientific personnel and put in jeopardy the very existence of the ITRI (a few, in fact, have closed down in recent years). The result is a vicious circle: ITRIs lose revenue and as a consequence the quality of their services goes down, which makes it more difficult to maintain their clients, and generates less revenue, etc. This is leading to serious doubts about the sustainability of these institutions, at a time when they could play a significant role in the process of competitive restructuring, by helping to upgrade the capabilities of productive enterprises to compete successfully in the world markets.

The ITRI Project

The "Industrial Technology Research Institutes Project", or ITRI Project for short, was jointly sponsored by. United Nations Industrial Development Organization (UNIDO) and International Development Research Council (IDRC). Its ultimate objective was to find out how to make ITRIs more effective and useful in the new context. Case studies were made of eight ITRIs in Latin America, surveying their role, activities and organizational characteristics (see Annexure 1). A common set of guidelines was used to help comparability. Case studies in Canada and New Zealand were also made as contrast.

Field-work took place during the second half of 1992, and the results were reviewed at a workshop in July, 1993. *Annexure 2* lists the reports prepared on the different case studies and offers brief comments on them. Summaries of the reports in the English language were prepared by the present author, and included in his 1993 report to UNIDO as part of a "Casebook". Additionally, the author conducted a series of interviews and examined written material from ITRIs and similar institutions in other parts of the work, incorporating the results as an additional series of "Cases" in the "Casebook".

Some Major Findings of the ITRI Project

The case studies could not investigate too closely the full range of questions originally posed, on account of limitations of time and resources available for each case study. On the other hand, the ITRIs studied are far from being a homogeneous set; they strongly differ in their particular milieu, history, fields of work, type of clientele, and so on. The peculiarities of the different cases made it difficult to get away from their idiosyncratic characteristics, and to a large extent each case is a special case; thus, it is not easy to draw conclusions and recommendations of general applicability. However, some aspects come out clearly enough and are presented below (we have incorporated some comments made by F. Machado in his study of July, 1993).

1. Evolution Towards Maturity

Maturity is achieved as a result of a long developmental process, as the ITRI solves its growth problems and finally achieves a "steady state" in which it fulfils efficiently a large proportion of the long-term objectives for which it was created, or which it has come to adopt. This may take many years. The evolution may be traced as an S-curve, time being plotted against the degree of fulfilment of the institute's "permanent" objectives Fig. 1.

During the developmental period the institute's main efforts will be devoted to solving the problems of growth and development, and the danger is that it will be arrested in this evolution, or worse still, that the curve will turn downwards because of lack of support, political difficulties, and other reasons.

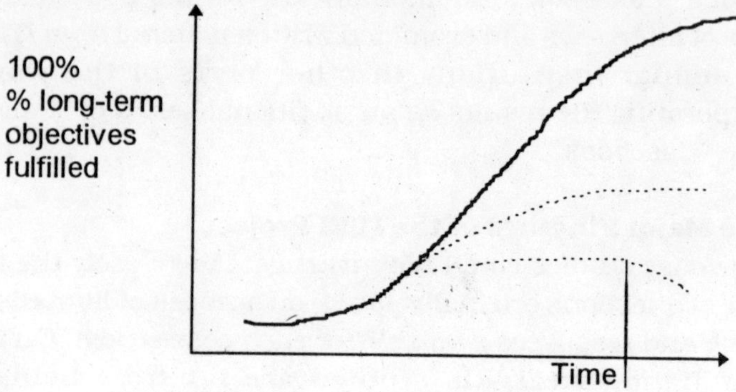

Fig. 1. Evolution of an ITRI.

Of the eight Latin American ITRIs studied in this project, only three may be said to have reached a satisfactory stage of "Maturity": Chile Foundation, CODETEC and Instituto de Investigaciones Technologieas (INTEC). In two cases there has been a sharp retrogression and the ITRI has ceased to exist Instituto de Investigaciones Tecnologicas (IIT), Colombia, and Instituto de Investigacion Tecologicas (IIT), Colombia, and Institutede Investigaion Technologica Industrialy Norrnas Tecnicas IITINTEL), Peru).

The three more successful institutes have evolved to deal with changes in their environment and their clientele, changing their structure and organization, selecting and nurturing their market niches, offering new services, and marshalling sufficient resources from their clients and from the State. In their flexibility and their ability to respond to challenges they differ clearly from the rest of the ITRIs, which are tied down to static, largely obsolete visions, concentrating on R&D projects which have frequently not found application, suffering from the decrease in State support, seeing their capabilities erode, and sometimes being at the mercy of political changes.

If an ITRI is going to be revitalized and improved, it is necessary to examine the validity of the long-term objectives in the new situation faced by the institution in the 1990s. A non-mature ITRI would require much more time and effort than a mature one, since it would be necessary to help it to develop towards a steady state where it can become fully effective.

2. *Role and Objectives*

As remarked by Machado, most of the ITRIs surveyed have not clearly defined their missions and have carried out little strategic planning. This may be due to several reasons, such as high political instability, lack of training and experience in management on the part of former scientists leading the institution, a rigid adherence to slow changing government guidelines, and expectations of continuing government support.

ITRI objectives appeared to be similar in most of the cases studied, and centred on providing industry with research results and other technological inputs needed. Research and development activities are mainly concerned with the use of local natural resources, the development of new products and minor innovations. Themes are frequently chosen by researchers, with poor transfer of results to industry. (This used to be the case in New Zealand and led to a strong restructuring.) Efforts have been too disperse, except for the Chile Foundation, CODETEC and CIMM; INTEC is now trying to concentrate its activities in a few major areas.

Emphasis on technology acquisition and diffusion have not been a major concern save in the case of the Chile Foundation. A proposal was made for CIMM to develop technology around which an enterprise may be created.

Because of the new environment and new technological conditions, enterprises need new types of services, such as constancy in a wide range of areas (strategic planning, technology management, organizational structure and culture, application of informatics); assistance with transfer of technology in general (search, evaluation and selection of technological suppliers, negotiation of contracts, technology brokerage); help in setting up strategic alliances and accessing technology through them; total quality services; pre-competitive and multi-client technology projects; rental of pilot installations, etc. These have been understood by the three better institutes, which have looked into the evolution of technological needs in the branches they cover, and have changed their service mix to attend to this evolution, going towards integrated solutions rather than isolated R&D and services. Other institutes are starting to explore new services: for instance, it has been suggested that CIMM should move into quality, automation and industrial waste.

The public sector has been a good client of the Latin American institutes in the past. However, public enterprises are being privatized and this takes away some important customers. The industrial clients of the institutes are mostly small and medium enterprises with relatively simple and limited demands.

The author was able to piece together some suggestions about how the role and objectives may be redefined:

- concentrate efforts instead of dispersing them among many lines of work;
- carefully identify clients and their demands;
- aim R&D to concrete needs of clients, and carry out long-term strategic research only when funding is available;
- provide technological services according to demand, but concentrate on those that industry or service firms cannot provide;
- give more support for technology acquisition, use and diffusion;
- pay strict attention to the generation of income through the sale of R&D and services;
- look for a balance between R&D and services, since too much of the latter may demoralize the best scientists and lower the quality of work in the institution.

3. Legal Regime

A public sector ITRI does not find it easy to be effective, for a number of reasons: there is a "public sector" mentality which is not favourable to efficiency; it is difficult to retain the best people, while the worst people cannot be fired; foreign exchange is hard to obtain, which affects the purchase of equipment and publications, the subscription to international information and communications systems, the travel of personnel, etc., there are legal obstacles to sell technology to private firms, or enter into joint ventures with them; and last but not least, political events may mean sudden changes in the ITRI's top management, as has indeed happened in several of the cases studied.

Two of the institutes reviewed, IIT and ININTEC, have faced a vicious circle as their funding shrunk: the best people left, those remaining lost motivation, the clientele became smaller, and as a consequence the financial crisis deepened, leading to further losses of top personnel, and so on. Such a situation finally brought about strong labour conflicts and led to the closing down of both institutes. CIEPE is running a similar danger.

Some way out of the rigidities of a public sector must be found if an ITRI is to be efficient. This may be a special regime giving it flexibility and commercial powers (New Zealand) or its transformation into a private enterprise or non-profit organization. However, as in the case of INTEC and of some instances in New Zealand and Canada, the State will very likely have to continue providing financial support to the ITRI for long-term research and for investments in new facilities.

4. Relations with Industry

The main shortcoming of Latin American ITRIs, with the exceptions of CODETEC, INTEC and Chile Foundation, is the difficulty they have found in forging good links with the production sector, and transferring them its technologies developed or acquired by the ITRIs. This distancing from what should be the principal client is marked by a lack of surveys regarding the relevance and the performance of the institute.

Some of the reasons for this would seem to be:

- enterprises show little interest in ITRI services, as a result of a long involvement with import-substitution industrialization; perhaps this will now change as firms find they compete much beyond a formerly protected national market;
- potential clients are principally small and medium enterprises with relatively low technological levels and

most lack science-minded personnel able to interact fruitfully with an ITRI;
- staff members of ITRI, particularly researchers, have often shown unhelpful attitudes, being more interested in what they perceive is "their own work" than in what industry needs;
- ITRIs have taken a long time to realize that in order to commercialize their services they require specialized skills and a marketing unit;
- there is a lack of policies which promote the demand for ITRI services.

5. *The environment in which ITRIs perform*

The milieu has not been stable in Latin America, with strong and sometimes sudden political and economic changes. The economic crisis of the 1980s was followed by a change in the economic model away from import substitution and towards trade liberalization; the retreat of the state from economic activity meant important cuts in subsidies and the privatization of public enterprises. The policies in favour of industrial restructuring and industrial competitiveness are still unclear and have not yet consolidated.

All this has meant a fundamental change in the environment in which ITRIs work, and is making life difficult for ITRIs. Two key issues are the curtailment of State support and the loss of some important State customers (as in the case of CODETEC). At the same time ITRIs are being pushed by governments to become more self-sufficient.

The goal of self-sufficiency is not easy to attain in the conditions under which ITRIs operate at present in Latin America. ITRIs are generally emphasizing the sale of services, mainly of the routine type, at unsubsidized prices that should at least cover costs. However, most ITRIs are not prepared for efficient marketing, often lacking necessary autonomy and flexibility, and do not count with up-to-date human resources and equipment. Additionally, the demand for ITRI services has decreased on the part of private sector firms, in particular small and medium enterprises, on account of the difficult situation faced by industry as the economy has been opened up and foreign goods have flowed in. This demand may not grow for some time, and in the mean time State support may still be needed to keep ITRIs alive, let alone to help them acquire new plant and equipment.

There is in Latin America a scarcity of policies aimed at promoting and increasing the demand for the services of ITRIs. The usual incentives to technological innovation in enterprises, such as outright subsidies, tax benefits, subsidized or risk-sharing credit schemes, supply of risk capital, government purchases of technologically demanding products, and so on, may not push enterprises to entrust their innovation efforts to ITRIs, given their tendency to use foreign technology and technical services. To do so would probably require a new breed of incentive schemes, such as are being tried now in New Zealand.

Continuing State support for research and technology institutions is a feature in Canada and New Zealand, the two countries on which comparison case studies were undertaken. In Canada, privatized ITRIs that expected to do away with State subsidies are again asking for support, unable to generate enough income from the sales of services to the private sector. In New Zealand, though, there is a decided effort to increase substantially sales of services by the new Crown Research Institutes, the latter's most important source of income still is, and will continue to be for some time, the "public good science fund", through contestable funding of 'strategic' R&D projects. This leads us to the point that State support doesn't necessarily mean budgetary allocations, promotional schemes for industrial firms to use ITRI services, and the creation of contestable funds for research and technology development in the public interest, as in New Zealand and more recently in Chile.

6. Resources

Financial resources: Latin American ITRIs have suffered constantly from shortages of funds. The State has been and still is the main source of funds, mostly through direct contributions, and more recently through contracts for research, as in Chile and New Zealand. International cooperation has been very important in getting started some ITRIs, and it is still crucial for the maintenance of ICAITI. The income from exploitation of technology has been negligible, except for the Chile Foundation, but some of the ITRIs have been successful in increasing sales of R&D and services. However, as we have pointed out above, State support is in general still needed: save for the Chile Foundation, which benefits from a large endowment fund, and the ITRIs studied do not appear to be sustainable without further State support.

Human resources: The training and improvement of staff have often been poorly planned and executed, there being (save for the Chile Foundation) scarcely any strategies regarding human resources. At present training is principally needed in new scientific areas and in marketing and management skills. Little attention is paid to keeping the staff up- to- date in different ways, such as the assistance to attend international meetings (only the Chile Foundation does it actively). Staff motivation is good only in a few cases, and there are almost no incentives based on merits/results; personnel evaluation systems are absent, or not applied; no exchanges of personnel with industry appear to take place. The use of outside consultants to supplement in-house skills is not usual, except for the Chile Foundation. Financial shortages have led to the loss of top people, have hindered training, and have led to recruiting new staff fresh from the university, thus lowering standards.

Physical resources and library resources: in general, they are outdated and it is still not easy to bring them to the level needed to cope with new technologies and to open new fields of work.

7. *Management*

The legal regime and structural characteristics of most Latin American ITRIs are not adequate for efficiency and impact. Organizationally we find rigid, pyramidal structures. Matrix structures seems to be little used, except for CODETEC. Additionally, as Machado points out, the attitudes, values and modes of behaviour that make up the internal culture were not satisfactory in most of the institutes- only the Chile Foundation presented concrete values of orientation towards the market, pragmatism, autonomy, innovation in services and models of transfer of technology. A delicate balance is needed between "working for the market," which should be the culture of an industry-oriented institute, and "working for science," which is what keeps scientists happy.

The main management functions of planning, programming, budgeting and evaluation could in general be vastly improved through the introduction of adequate management techniques (like in the Chile Foundation) and the training of scientists who have taken up managerial positions. Human Resource Management, a crucial aspect, is in dire need of improvement, as indicated above.

Few of the ITRIs surveyed had clear ideas of the characteristics, size and likely evolution of their markets, of what competition there was or was likely to be, of what new markets could be tapped, what alliances were of promise, and so on. It was clear that in most cases the marketing function is in need of improvement, if ITRIs are to become more self-sufficient. This would include training the staff in sales techniques, analyzing the existing and potential clientele and promoting the institution through the media and through presentations in local and international forums.

Machado stresses the need for strategic management and strategic planning as a means to continuously steer institutions towards excellence, survival and growth. An ITRI, he feels, should learn about impending changes in its environment (technological trends, characteristics of their clients and the evolution of their likely technology needs), identify new markets and areas of work, choose the best research areas (strategic and speculative), etc. On the basis of such an analysis it should take the necessary anticipatory decisions regarding market strategy, service mix, the development of critical capabilities, changes in its structure and operating systems, etc., to allow for a quick response to the needs of clients, the development of its human resources and the fulfilment of its "development vision".

The case study in New Zealand shows how research institutions in the public sector (the new Crown Research Institutes) and in the private sector (the Industrial Research Associations) pay great attention to these aspects, drawing up five-year strategic plans and two-year business plans, an example that Latin American ITRIs should examine carefully.

8. Outputs and performance

"Accountability" is not yet a significant characteristic of Latin American ITRIs. Some of them do not seem to have any formal systems to control closely the cost, timeliness, quality and relevance of their work.

Research work carried out in the ITRIs has rarely led to a patent, and the production of "technology packages" as an end result of R&D is not yet common. The main products of ITRIs are technological services of varying degrees of complexity. While the selection of areas of work and of R&D topics is gradually being oriented towards the market rather than towards the interests of scientists, much remains to be done to achieve a demand-pull situation.

Save for two cases (the Chile Foundation and to a lesser degree CODETEC), there appeared to be a limited involvement of the ITRIs in technology acquisition from foreign sources, and in the diffusion of such technology, though, these activities have been the most important source of technological innovation in the past for industry. The example of the Chile Foundation is noteworthy in that one of its principal activities is to identify promising foreign technologies that may be used in Chile in its areas of concentration (agro-industry, marine resources, forestry and wood products), bring them into the country, and apply them in "demonstration projects" in which it invests its own resources. After a few years the investment is sold and the cycle starts again. About 30 new firms have been started in this way, some with the support of private investors. Several have become a starting point for new productive activities in Chile, as industrialists have invested in new production facilities, with the help of abundant information and technical support from the Foundation. The case of salmon production is probably the best example: a whole new agro-industrial branch came into existence after the Foundation showed the way with its own firm, Salmones Antartica SA, and ten years later the new activity was exporting fish over US $200 million a year.

Another remark under this heading is that Latin American ITRIs tend to work on their own, and in general do not explore possible complementarities with other providers of research and technological services. They should look into the possibility of increasing considerably their links and their cooperation activities with other local scientific and technological institutions, which may contribute to the consolidation of a national innovation system.

As we have indicated above, it is to be expected that in future years the ITRIs will pursue a number of activities that the new situation increasingly requires, such as the production of technology packages for clients, consultancy services in technology management, productivity, quality, new services related to environmental questions, creation of new technology-based firms, foreign technology acquisition and assimilation, and assistance to local enterprises to enter into joint ventures and strategic alliances with foreign, technologically dynamic firms, an area of growing importance if Latin American enterprises are to play a significant role in advanced technology fields.

ITRI Revitalization

An important conclusion of the ITRI Project was that several of the Latin American institutes covered by this study were not carrying out appropriately the roles they should play in the new industrial and economic environment, and were in need of change and revitalization to enable them to carry out efficiently such new roles.

The author feels that this conclusion also applies to other ITRIs, in Latin America and in other developing regions of the world. Should this be confirmed, we would find that a good number of ITRIs in developing countries (and in the former socialist countries, where conditions have changed drastically in recent years) are still far from being effective. It is of great interest to consider what revitalization efforts could be undertaken to make them so.

By "revitalization" we understand the process of transforming an existing ITRI into an effective institute. This needs not only the improvement and restructuring of the institution to assure efficient operation, but also a redefinition of its role and objectives to enable it to have a high impact on development.

1. An effective ITRI

A high degree of effectiveness in an ITRI may be said to have two components. First, high efficiency regarding the way operations are carried out and outputs are transmitted to the users. Second, the achievement of a satisfactory impact in favour of the country's goals, taking into account the new environment of the 1990s.

There is no easy way to quantify these parameters, with the exception perhaps of some aspects of efficiency, but expert observers will in general be able to pass a qualitative judgment on them.

Though the notion of what constitutes an effective ITRI may vary from country to country, it is possible to draw up a list of aspects that would seem to be shared by many effective ITRIs:

(a) ***General features.*** The ITRI should have achieved maturity, including a size which assures a minimum efficient "critical mass". Its legal status should assure flexibility in its operations (allowing all types of commercial operations) and in the management of personnel. It should have explicit, clear objectives, in line with the country's goals and with the needs of industry. It should avoid the dispersion of efforts, choosing a limited number of technological areas and types of services, including R&D, to be offered to clients.

(b) ***Finances.*** The ITRI should count with stable, adequate funding, a high proportion of which should originate in sales to its clients, and in the commercial exploitation of the technology it has developed or introduced into the country.

(c) ***Resources.*** The ITRI should count with a cadre of capable scientists and engineers, well trained and highly motivated. This staff should be managed carefully in the different aspects of recruitment, training and updating, and evaluation. Annual staff turnover should be low, reflecting favourable conditions regarding work satisfaction and remuneration. The access of scientists to world science and technology should be encouraged, particularly through exchange programmes and the assistance to international seminars and meetings.

The institute needs adequate physical installations in buildings, equipment, instruments, computing, communications, and so on, to be regularly updated in line with scientific and technical progress. It also needs adequate library and information facilities, including subscriptions to reviews and fluid connection to outside information sources and databases.

(d) ***Management.*** The institute should have sound, modern management practices, including a long-term strategic plan and a short-term (2-3 years) business plan. Most, if not all, work undertaken in research and services should be organized as projects for which funding should be assured at the outset. Projects should be managed carefully both at the design and execution stages. Institute managers should be trained in management techniques.

(e) ***Outputs.*** R&D topics should be carefully chosen. In the case of strategic research, choice should be made according to the relevance to long-term national needs. In the case of short-term research, it should closely follow client demand, ascertained through close contacts with clients and familiarization with their needs, and there should be an effort to produce "technological packages" of direct applicability.

The product mix of technological services should be dictated by the market. There should be a periodic review of the outputs, suppressing those for which there is little demand and creating new types of services for which demand is growing.

The institute should assure the quality and timeliness of the research and services produced.

(f) ***Clients and marketing.*** The institute should forge strong, stable links with industry, through different means such as industry membership in the board of directors and advisory committees, the formation of consortia with industry, periodic visits, the exchange of personnel, public relations efforts, the economic analysis of the client sectors, etc. The marketing of the ITRI's outputs should be a primary task, which may be entrusted to a special unit closely linked to the technical units. The institute should fully understand the characteristics and problems of the industrial branch or branches it is serving, and to this end should carry out economic and market studies.

Some Examples of Effective Institutes

A number of ITRIs in developing countries have been able to carry out the transition from an "R&D laboratory" to an effective ITRI. Among the cases reviewed in Latin America, only the Chile Foundation, CODETEC and (perhaps) INTEC have been able to do so. There are interesting examples of other developing country ITRIs that have achieved a good degree of effectiveness, such as Taiwan's Industrial and Technological Research Institute, India's National Chemical Laboratory, Korea's KIST and Argentina's INVAP, to name a few.

Of the institutes surveyed in the ITRI Project, we would like to single out two interesting examples of reasonably effective ITRIs:

- The Chile Foundation, which has defined carefully its objectives and its field of action, adopted a legal figure and a managerial structure that allow it to conduct its operations with high efficiency and flexibility, established excellent relations with the productive sector in its areas of concentration, and achieved a high degree of maturity, including self-financing;
- The New Zealand Forest Research Institute, which covers the whole cycle that goes from the genetics of relevant tree species to the exploitation of forests, the industrialization of wood and the commercialization of wood products. It has developed a strong scientific base and produced significant applied R&D findings within its fields of specialization. At the same time it has managed to build up excellent relations with the different productive sectors it attends to, and to service them with research and technological services. The recent change in its status, from a traditional public sector science organization to a Crown Research Institute, has endowed it with flexibility and commercial powers that are likely to allow it to become even more relevant to the important wood products sector in the country. FRI is not yet self-financing; it enjoys a significant support from the state, mainly through the funding of strategic research by the contestable Public Good Science Fund, but the

generation of income from sales to customers of its R&D and services has been on the increase. It should be noted finally that FRI has been in many ways a blueprint for the concept of Crown Research Institutes, a set of effective S&T institutions created in the recent reform of New Zealand's public science sector, in lieu of the unsatisfactory science units formerly in existence.

Aspects in the Revitalization of an ITRI

When an ITRI has not managed to achieve maturity after many years of operation, and is still struggling with its own development process, the problem of helping it along and revitalizing it is not just a question of correcting different aspects and adjusting to new environmental characteristics. The ITRI would require assistance to complete its development through the acquisition of know-how, the development of human resources, increase in the efficiency of management, creation of closer ties with industry, and other aspects. At the same time an eye should be kept on the new needs stemming from the current developmental paradigm, the new policy environment and the increasing scarcity of State funding, all of which would call for a change in the original objectives established at the time of foundation. All this would require further investments, which would not seem possible without adequate financing from the State or another external source such as international cooperation.

Sometimes the situation may have deteriorated to a point where it may not be feasible to revitalize the institute. Three Latin American ITRIs have disappeared in recent years, in spite of strenuous efforts to keep them alive, and others are in danger of doing so. When such retrogression has set in and "vicious circles" have been operating for some time, the best alternative may perhaps be to close down the institute and create a new one with the best design and conditions to make it effective.

Some ITRIs have conducted strategic planning exercises, trying to peer into the future and designing strategies, policies and actions for the institution. Most of these exercises, however, usually accept the main structural characteristics of the institute, and consider most government policies as given. A revitalization exercise should go much further. It should explicitly aim at introducing desirable changes in all aspects of the ITRI's structure and its legal status, and should make explicit proposals to the authorities regarding new policies, as well as changes in the existing policies that will help the institute to increase its impacts and, thus, become fully effective.

How a particular ITRI may be revitalized will depend on a variety of factors, such as the characteristics of the national economy and of the policy environment, the peculiar history of the ITRI, its size, its chosen fields of concentration, the peculiarities of its clientele, the support it can get from the State, the industrial environment in which it operates, the limitations to which it must adhere, etc.

The revitalization of an ITRI needs attention to many interrelated aspects. Principal among them would be the policy environment for industrial technology development; the restructuring and improvement of the ITRI itself; cooperation of the ITRI with industry; cooperation of the ITRI with other institutions in the country, principally S&T centres and universities; foreign and international cooperation, and scientific and technical cooperation with foreign ITRIs and S&T institutions.

Examples

Case studies carried out in the ITRIs Project show attempts that have been made to change and restructure some Latin American institutes. This has happened with CODETEC and INTEC when a change in the policy environment eliminated some important customers, lowered government support, and brought about demands for different R&D and technological services. In both cases reforms were introduced and the outcome seems to have been a happy one. In the case of ICAITI a strategic planning exercise has made recommendations which, when put in practice, will affect many aspects of the institute's operation, though, not its main structural characteristics. The case of CIMM is of interest too since the researcher, together with the senior staff, has advanced a proposal for a deep change in the legal framework (privatization), the fields of concentration (a new emphasis in what industry is now needing: R&D and services in automation, quality, waste processing) and the notion of "producing" technology-based enterprises. This proposal may be the basis for a thoroughgoing CIMM revitalization exercise at some time in the future.

The most important example of restructuring reviewed in the ITRI Project took place in New Zealand (case studied by the present author). This country has undergone a series of changes and adjustments in recent years, as it became clear that the economic model previously followed by the country, with a largely protected economy relying on the export of agricultural commodities, did no longer generate enough wealth to maintain reasonably full employment and income growth. The restructuring of science and technology in the public sector is one of the most interesting chapters of such changes. The functions of S&T policy-making, R&D funding and R&D execution were separated. A fund was created to support "public good" strategic R&D. The four main science agencies were restructured: ten new Crown Research Institutes took their place. CRIs focused on sectors and problem areas rather than disciplines, and have been endowed with full autonomy and commercial powers in the hopes of linking them closely to the productive sector.

Conducting a Revitalization Exercise

To help in the revitalization exercise we have prepared the attached Checklist (Annexure 3). This may be used as a guide for examining the many areas where action may have to be taken, or induced, to revitalize an ITRI. The different topics are ordered by the main subject. For each point there should be an examination of the present situation and an indication of the desirable evolution towards the revitalization of the institute.

It should be noted that this checklist does not intend to establish a sequential or linear procedure for the revitalization exercise. There is a strong interrelationship among many of its points, and, therefore, the procedure should be one of successive iterations. The checklist may also be useful when a new Industrial Technology Research Institute is being designed.

The following steps may be suggested:

(i) *Diagnosis of initial conditions:* The ITRI's strengths and shortcomings, changes taking place in its environment, possibilities for making the ITRI effective. The checklist may be used in this and the following step.

(ii) *Design of the revitalized institute in its desirable new steady state:* Long-term objectives, structure, outputs, clients, resources, and so on.

(iii) *Design of the transition towards the new steady state:* Sequences and time schedules should be established, taking into account the need to enact new policies on the part of government, change the legal status and the organizational structure, discontinue inadequate staff, hire new personnel, train scientists in new fields, impart marketing and management training, carry out necessary investments in buildings, equipment and other assets, and so on. The costs of these items should be carefully estimated and the required funding should be procured from national and international sources.

(iv) *Implementation of the revitalization proposals:* This needs continuing support from the ITRI's sponsors, as well as from government, industry and the science system, since the transition to a new steady state will follow an "S" curve, and cannot be overly accelerated; the ITRI can only start producing the expected outputs and results when people, installations and proper funding are in place, the new techniques have been mastered, and the clientele has materialized.

(v) *Organizing the revitalization exercise:* This may be done through a special planning group or task force, including some top foreign experts. Ample consultations should be held at different moments with the stake holders in order to build a broad consensus that will widen support.

Annexure 1: Case Studies in the ITRI Project

The institutes studied were:

— Argentina: Centro de Investigacion de Materiales y Metrologia (CIMM);

— Brazil: Companhia de Desenvolvirniento Tecnologico (CODIZEC);
— Chile: Instituto de Investigaciones Tecnologicas (INTEC);
— Chile: Fundacion Chile;
— Peru: Instituto de Investigacion Tecnologica Industrial y Norrnas Tecnicas (ITINTEC);
— Colombia: Instituto de Investigaciones Tecnologicas (IIT);
— Venezuela: Centro de Investigaciones del Estado para la Produccion Experimental Agropecuaria (CIEPE);
— Central American Region: Instituto Centro Americano de Investigacion Tecnologia Industrial, ICAITI (Guatemala).

Additional Cases:

- *Canada:* Some recent initiatives in the development of industrial technology
- *New Zealand:* Restructuring of the public sector S&T system and role of the new Crown Research Institutes.

The methodology employed was based to a large degree on the proposals made by D. Chudnovsky and R. Bisang in a paper commissioned by UNIDO. The project pursued a number of specific objectives, principally: — How to facilitate the restructuring and transition of ITRIs for effective work in a new environment, with a capability to respond to the technological requirements of industrial firms; — How to assure the sustainability of ITRIs in the face of reduced government funding; — How to make ITRIs more effective, and how to balance demand-driven work with long-term work oriented to future requirements.

Annexure 2: Papers Prepared for the ITRI Project

(*a*) ***Reports on the Case-Studies*** (followed by a short commentary).

Araoz, A., 'Restructuring of science and technology and new modalities of industrial research in New Zealand', Cambridge, Massachusetts, March, 1993.

New Zealand has undergone a series of changes and adjustments in recent years, as it became clear that the economic model previously followed by the country, with a largely protected economy relying on the export of agricultural commodities, did no longer generate enough wealth to maintain reasonably full employment and income growth. The restructuring of science and technology in the public sector is one of the most interesting chapters of such changes. The functions of S&T policy-making, R&D funding and R&D execution were separated. A fund was created to support "public good" strategic R&D. The four main science agencies were restructured: ten new Crown Research Institutes took their place. They are focused on sectors and problem areas rather than disciplines, and have been endowed with full autonomy and commercial powers in the hopes of linking them closely to the productive sector.

Cordua, J., Rol de los institutos tecnologicos industriales en el nuevo contexto social y economico mulldial: estudio de casos de Fundacion Chile," Fundacion Chile, Santiago, February, 1993.

The Chile Foundation offers a very interesting contrast to the usual industrial technology research institution in a developing country to the usual industrial technology research institution in a developing country. It is self-sustaining, shows a strong commercial outlook, hardly makes research, and had a strong impact on the productive activities of the country in key export sectors, through the acquisition of foreign technology which it has incorporated into "demonstration projects".

Flit, I, "EI INTITEC del Peru : un drama technololgico," Lima, January, 1993. The institute was closed down even as the case study took place, after a long period during which it suffered many ups and downs for different reasons, finally losing the support of the authorities. It is important to ask how and why this happened, and why industry did not fight to salvage the only general purpose ITRI in the country. These and other questions are explored in this case study.

Jaramillo, L.J.,"la larga crisis del IIT: difcultades para una capacidad technologica industrial en Colombia", Bogota, February, 1993.

The Institute of Technology Research of Colombia, IIT, was closed down recently and its assets were dispersed, marking the unfortunate end of a long crisis. As in the previous case, the author analyses the how and why of this demise.

Machado, F."EL Instiuto Centroamericana de Investigation Y Technologia (ICAITI)" San Jose, Costa Rica, February, 1993.

This case shows an ITRI with the status of a subregional institution, conceived to serve the industrialization process of the five Central American countries, which 30 years ago agreed to form a Common Market. The characteristics of Central American Industry — small scale, technologically backward, oriented towards the local markets, demanding relatively simple services and hardly able to pay for them — have not made life easy for ICAITI, which has relied to a large extent on foreign and international cooperation to build up its capabilities and carry out research activities for its member countries. At present ICAITI has to find ways and means to improve its human, physical and financial resources, its efficiency, and its link with industry.

Mullin, J., "Some recent Canadian initiatives in the development of industrial technology", Ottawa, February, 1993.

This study reviews some institutional policies an innovations in Canada that may provide some possible future directions useful to ITRIs in developing countries. It analyzes the "technology policy environment" in Canada today and presents four cases studied.

Soto Kerbs, L., "Evaluation del Instituto Technologico de Chile", Santiago, January, 1993.

This is the case of an ITRI in Chile which has effected a successful transition of the present "liberal" setting. Chile was the first country in Latin America to abandon the import substitution model and open up its economy. This happened ten years ago, and eventually led to a sharp increase in export and a high and steady rate of economic growth. Adapting to the new model was not easy for INTEC, and the way it was done has interesting lessons for other developing countries.

Sauza Neto, J.A and Jose Vidal Bellinetti, "Case Study of CODETEC", Sao Paulo, February, 1993.

This institute, born within a top Brazilian university and later evolving largely on its own, has undergone a series of adaptations and transformations as it went from a liaison mechanism for companies university to an instrument of Brazil's import-substitution policy in the field of pharmaceuticals to an ITRI that, keeping to the technological field where it has built strengths, is trying to survive and prosper in the new liberal economic regime installed in 1990

Validivieso, R., "Estudio de caso: Fundacion CIEPE", Caracas, February, 1993.

The CIEPE Foundation of Venezuela was created to support industrial development in its region, mainly relating to agro-industry. The institute never had a chance to develop properly, principally because of the lack of resources and the influence of politics and changing policies. At the time of the study it was in a very precarious position, with strong budgetary problems and labour conflicts.

Varotto, C., "Hacia un nuevo rol para los Institutos de Tecnologia Industrial en los paises en desarrollo: estudio del Centro de Investigacion de Materiales y Metrologia," Buenos Aires, January, 1993.

ClMM is a relatively small public sector ITRI in Argentina, founded in 1963, which has been able to make only limited progress, despite the stability of its management and strenuous efforts to serve local industrial firms. The author of the case study has made a careful analysis of CIMM and has set forth a number of suggestions for restructuring and revitalizing it under the new policy environment the country has recently adopted.

(b) ***Other Papers***

Araoz, A., "The Revitalization of Industrial Technology Research Institutes in Developing Countries. Guidelines for UNIDO", a report to UNIDO, Vienna, 1993. This report presents the results of the ITRI Project and a proposal for ITRI revitalization. It includes a "Casebook" with summaries of the case studies. A condensed version was made available at the 1995 WAITRO Conference.

Chudnovsky, D. and R. Bisang, "Towards a new role for ITRIs in developing countries," Buenos Aires, March, 1992. The Original Guidelines for the case studies.

Machado, F., "Revision de literature sobre la vinculacion Institutos de Investigacion Industrial/Firmas Industriales en

America Latina, San Jose, Costa Rica, Jan. 1993. A review of literature on the linkages of ITRIs and industrial firms in Latin America.

Machado, F. "Institutos de Investigacion Industrial en America.

Latina: su rol en los anos noventa. Sintesis de los estudios de caso, San Jose, Costa Rica, July, 1993. A synthesis of the case studies.

Annexure 3: A Checklist for ITRI Revitalization

1. Environment in which the ITRI works

- Government policies and their changes under the influence of new international techno-economic conditions (fast technological progress, global competition, globalization); this would be expressed in short-run stabilization policies, opening up of the economy, and policies for higher productivity and competitiveness;
- Stability of government policies and programmes;
- Influence of political and economic upheavals, such as war, sudden price changes of raw materials in the international market, internal violence, etc. (this marks a strong difference between, say, Denmark and Jordan).
- Obstacles such as difficulties in importing publications, equipment, reactives, and supplies;
  Existence of a "development infrastructure helping the ITRIs to interrelate with government, industry, and other science and technology institutions;
- Explicit government policies in favour of technological development in industry: (a) existing measures, and (b) Measures that may be proposed.

2. Characteristics of the ITRI

- ownership and legal regime of the ITRI;
- present ITRI objectives, formal and actual; how do they differ from the original objectives ?
- maturity of the ITRI;
- productive sector or sectors towards which it is aimed; technological areas it covers;
- clients;
- services.

3. Resources

3.1. Financial resources

- origin;
- government, through a budgetary allocation, grants, contracts for strategic R&D;
- industry, through membership fees, sales of multiclient services, including R&D, applicable to a group of companies, sales of services including R&D, to individual companies;
- sales of services, including R&D, to other clients, such as government agencies, other SST institutions, etc.
- sales to clients outside the country;
- income from proprietary technology developed or procured by the ITRI, through sale, licensing or joint ventures;
- adequacy of funding: ample, barely sufficient or subcritical.

3.2. Human resources

- structure of personnel;
- level of scientific and technical staff;
- availability and utilisation of outside consultants.

3.3. Physical resources

- Institutions, buildings, equipment, instruments;
- Computer system, Communication network.

3.4. Knowledge and Information resources

(A) *For scientific and technological purposes*

- Documentation.
- Library.
- Reception of S&T reviews, periodicals and publications.
- Connection to S&T information system.
- Scientific and technological intelligence activity.
- Personal access to international scientific and technical information.
- Assistance at technical meetings, locally and overseas.
- Training and specialization of scientific personnel particularly through stay overseas.
- Participation in collaborative research and other

activities with other local and foreign institutions–Participation in scientific networks.
- Use of outside consultants.
- Procurement of foreign commercial technology by purchasing, licensing and other means.

(*B*) *For Economic, Commercial, Regulatory and other purposes*
- Library.
- Reception of publications.
- Contacts with industry, government, academic circles, and the media.

4. Management

4.1 Structural characteristics
- Organizational structure.
- Flexibility: legal, actual.
- Level of management skills of persons in manage-ment positions.

4.2 Planning, programming, budgeting, evaluation
- Strategic planning.
- Annual business plans and production programmes.
- Programme formulation for R&D project and other types of activities.
- Budgeting, formation of reserves, internal allocation of resources.
- Efficiency with which programmes are implemented.
- Evaluation, ex-ante and ex-post.

4.3 Human resource management
- Recruitment.
- Training and upgrading, including assistance to scientific and technical events.
- Staff exchanges with industry.
- Adequacy of remuneration; salary incentives motivation and work discipline a "university" or a "production" setting.
- Evaluation of personnel performance.
- Degree to which key personnel are retained in the long-run.

- Recruitment and utilisation of outside consultants, national and foreign.

5. Outputs and performance

- Performance indicators : financial and other types.
- Performance of individual functions.
- Research and development.
- Type of R&D : basic, applied, adaptive, strategic, speculative, contract research for a client.
- How topic are chosen.
- Relevance to country's short and long-term needs.
- Relevance to customers' needs.
- Efficiency in performance of research.
- Quality of the research.
- Development of usable, commercial technology packages on the basic of research result.
- Transfer of technology to users.
- Technological services: information, consultancy, testing, metrology, extension, etc.
- Characteristics affecting efficiency and relevance.

6. Relations and cooperation with industry

- Industry participation in board and committees.
- Links with industry.
- Types of links.
- Barriers to demand for industry.
- Policies encouraging such demand.
- Behaviour of industrial firms : their technology strategy, their need for doing in-house research or using the ITRI labs, their involvement in planning, development and support of the ITRI.
- Involvement of ITRI in technology transfer to local companies from foreign sources.
- Industrial survey and marketing studies made by ITRI.
- Organization of consortia of enterprises for cooperative research.
- Image of ITRI in industry.
- Marketing by ITRI of its services.
- Licensing of technology developed by the ITRI.

7. Cooperation with other National Institutions

- Cooperation with other ITRIs.
- Cooperation with basic and applied research institutions which are not primarily concerned with industry.
- Cooperation with universities.
- Reinforcement of the "national innovation system."

8. International Cooperation

- Multilateral cooperation programmes, particularly with UNIDO and the UN system.
- Bilateral cooperation programmes.
- Cooperation with ITRIs in developed countries Twinning.
- Cooperation networks with ITRIs in other countries. Participation in WAITRO and in regional networks.

3

Cooperative R&D Institute: Status Report of South Korea

Chang Jae Lee   Deok Soon Yim

It is well known that technological innovation is often driven by competition. Firms set up strategies for technological innovation in response to competitors' strategies. Recently, however, strategies for technological innovation have come to encompass cooperative schemes as well as competitive tactics (Bozeman, 1991). These changes take place in the S&T environment. Firms begin to modify traditional technology strategy, intertwining the tactics of competition and cooperation. Typically, leading firms have formed "strategic alliances" among firms within the same industry to enlarge their capabilities.

National R&D programmes in high-technology fields have involved cooperation schemes for enhancing effectiveness and efficiency of Research and Engineering (RE) activities. Many mega-sized joint projects have recently evolved in developed countries. There are VLSI and IMS programmes in Japan, AIHSIC and SEMATECH programmes in the USA and EUREKA and ESPRIT programmes in the European Union (EU). Moreover, developed countries have promulgated various promotion laws in order to encourage cooperative research. For instance, the USA promulgated the Stevenson-Wydler Act in 1980, the National Cooperative Research Act in 1984 and the Federal Technology Transfer Act in 1986, all of which stipulate mechanisms of cooperative research.

Definition

Cooperative research and development R&D is a pattern of various strategies for technological development that utilizes the concept of cooperation. Cooperative RE or joint R&D can be defined as a joint effort of more than two R&D actors of the national innovation systems in order to develop and acquire new technologies. Currently, the term of "cooperative R&D among industry, universities and research institutes" (San-Hak-Yun cooperative research) is very widely used in Korea (Cho, 1992). In this term, "research institute" means government research institutes (GRIs) which have played an important role as mediators between the two traditional partners of cooperative research industry and universities.

In a broader sense the "industry-universities-research institutes cooperative R&D" means a variety of direct and intentional joint efforts or activities between industry and universities, between industry and GRIs, and among industry, universities and GRIs. The activities of this cooperative R&D include joint technology development, technology transfer, and human resource training through joint project operation.

Reasons for the Popularity of Cooperative R&D

Some theorists consider the research consortium of cooperative R&D as being the most appropriate technology management system in the knowledge-intensive society of the 21st century. In fact, in developed countries, cooperation between industry and the university has been practised for many years. This type of cooperation becomes a typical pattern of cooperative R&D. This conventional cooperation scheme and other types of cooperation are more highlighted now than in the past. There are several reasons for the re-emergence of the cooperation concept.

First, the distinction between basic and applied research and development has become increasingly less clear, and the commercial value of basic and applied research is sometimes very high. As a result, industry has begun to show an interest in basic and applied research, providing research funds to universities. Secondly, as projects of advanced and core technology cover multi-disciplinary research fields with a wide research scope, the R&D funding required for research is getting larger while, at the same time, the risk of the investment is getting higher. Accordingly, firms dealing with risky projects have changed their development strategies from individual to joint development with other firms (Friar and Horwitch, 1985; Cho, 1992). Thirdly, nations have reinforced their efforts to increase international competitiveness by utilizing their technological potential as well as S&T resources. As a means of enhancing national technological capabilities in high-technology industry, governments have launched cooperative R&D programmes at the national level.

Reasons for Promoting Cooperative R&D in Korea

For Korea, which lacks self-supporting technological capabilities and whose R&D resources are not equally distributed among the industry, universities, and government-funded research institutes (GRIs), the "concept of cooperation" is essential to enhance the efficiency of R&D investment. As the level of technology that Korean industry deals with is improved, adopting new tactics combining R&D resources of various innovating actors is very conducive to new product and process development. The cooperative scheme contributes not only to the enhancement of self-reporting technological capabilities but also to the promotion of international competitiveness in the advanced core technology fields. As research outcomes from GRIs or university laboratories have proved to be valuable sources for the technological innovation of industries, the Korean government encourages the tri-party cooperation and accelerates the establishment of the support system for the tripartite cooperation.

This paper examines: (1) the Korean cooperative R&D system and its current status, (2) major motivation of cooperative research and difficulties in implementing the system, and (3) the Cooperative R&D Promotion Law that was enacted in Dec. 1993. In conclusion the study presents cooperative R&D problems in Korea and possible solution for the future.

Need for Cooperative R&D

In Korea most of the industrial technology has been introduced through technology importation and imitative reverse-engineering. As a result, industry has not had to hold a large portion of R&D manpower nor extensive research facilities. Instead, universities and GRIs have held most of the high-calibre research manpower. This imbalance is an outcome of several factors. First, Korean firms have set up production facilities through international technology transfer or technology imitation rather than technology development. Thus, they did not have to hire a large number of researchers. On the other hand, the government established GRIs in the 1970s and the GRIs have actively hired many doctoral or master's degree holders. Secondly, Confucian cultural heritage which regards academia as a higher social status than industrialists, had hindered manpower movement from the university to industry until recently.

At present, industry accumulates manufacturing technological capabilities experiencing many difficulties in developing product and process. Industry spends enormous amounts of financial and human resources in R&D activities. R&D investment by industry amounts to 80 per cent of the total national R&D investment. Industry holds 59 per cent of the total research manpower. In terms of the quality of the research staff, however, they do not have sufficient human resources to pursue independent product and process development. Industry had only 8 per cent of the Ph.D. degree holders in Korea. On the other hand, Korean universities, holding 76 per cent of the Ph.D. degree holders, lack research facilities and R&D funds. They have emphasized education functions but have not actively participated in manufacturing technology development (Table 1).

As the international competitiveness of Korean products has weakened since the late 1980, i.e., government has been trying to maximize national R&D capabilities by combining the high-calibre manpower of universities and GRIs that have the product development capabilities of industry. The government's efforts to strengthen cooperative R&D are well represented in its policies and regulations. The main thrust of these efforts is to combine R&D resources of the three innovators, which could be complementary to each other. Through this combination, the government wants to link basic and applied research conducted by universities and by way of product development activities performed by industry and ultimately to the commercialization of technologies.

Table 1
Strengths and Weaknesses of Industry, Academia and Research Institute

	Strengths	*Weaknesses*
Industry	— 80 per cent of R&D investment, 59 per cent of R&D manpower product development know-how — Efficient manufacturing facilities	–8 per cent of Ph. Ds —Heavy dependence Technology import and simple modification of imported technology short-term development plan
University	— 76 per cent of Ph Ds — Accumulated know-how in basic research	—Emphasis on education and basic theory —shortage of facilities and fund — Luck of product development experience
Government funded Research Institute	— Excellent manpower and facilities — Abundant R&D experience — Well-established R&D system	— Shortage of field experience — Weak basis of manufacturing technology development system

Classification of Cooperative R&D

Classification by Innovative Actors

Cooperative R&D among industry, universities and GRIs in Korea can be classified into four types in terms of innovative actors who initiate cooperation. There are bureaucrat-initiated, GRI-initiated, industry-initiated and university-initiated joint research. The government bureaucrat-initiated cooperation is one in which bureaucrats function as managers and mobilize researchers of industrial, university, and government labs for predetermined projects. The GRI-initiated cooperation is mostly related to training technical manpower from industry and the university. Industries and universities send their engineers and technicians to GRIs where they learn advanced knowledge and absorb new technology. This type of cooperation takes the form of either small-funded research projects or training projects between GRIs and industry or universities. During the period of the small-funded project, a contracting firm may use the research facilities of the GRIs or receive consulting services from GRIs' researchers. The expenditures incurred for the use of the facilities and consulting services are calculated at the end of the project.

Industry-initiated cooperation has at least two underlying purposes. One is that a firm donates research money to universities for concrete, usable or marketable research results. The other is that in return for some donation from a private firm university professors are expected to recommend to the firm good students as new employees. The university-initiated model of joint research may be described as a type of fund-raising, whereby a university seeks new financial sources for support of its research programmes by establishing ties with the industry.

Since each has its own internal logic and self-survival aims, these four types can be seen as being independent from each other. It is difficult to judge which will prove to be promising for the future of science and technology in general, but it is certain that industry will increasingly take initiatives in the fields of basic and applied research as the level of firms' R&D increases.

Classification by Activity Patterns

Cooperative R&D among industry, university and research institutes can be classified into four categories in terms of activity patterns. Cooperative research is categorized into a research activity and an education activity. Each category is divided again into cooperation with manpower involvement and cooperation without manpower involvement.

The category of manpower cooperation in research activity includes joint research, projects commissioned by industry to the GRIs and university laboratories and technology consulting for the industry. The category of non-manpower cooperation in research activity includes co-use of research facilities and technologies, the exchange of technology information and technology transfer. The category of manpower cooperation in education activity includes the exchange of professors, researchers, and students and the participation of private companies in the training programmes of the universities. Finally, as for non-manpower cooperation in education activity, GRIs and private companies offer research facilities and scholarships to universities (Table 2).

Table 2
Activity Patterns of University, Industry and GRI Cooperation

Classification	*Strengths*	*Weaknesses*
Research Activity	- Joint research common goals - Project commissioned by firms to universities for GRIs - Technology consulting to firms-	Co-use of research facilities - Technology transfer - Exchange of technology information
Education Activity	- Exchange of professors, researchers - Exchange of students, field training at firm Researchers enrolling graduate courses - Industry's participation in training programmes at universities	- Offering research facilities to universties (donation or allowance of free use) - Exchange of training materials and information - Providing scholarship to firms, Strategy to secure high-calibre manpower.

Current Status of the Cooperative R&D

Bureaucrat-initiated Cooperation

Cooperative R&D has the first priority among other RS projects to receive research grants from government. There are several national programmes funded by the Korean government: the National R&D Project (NRP), which has been administered by the Ministry of Science and Technology since 1982 and the Industrial-base Technology Development Project and the Alternative Energy Development Project, which have been administered by the Ministry of Industry, Commerce and Energy since 1987 and 1986, respectively. For each of these national programmes, there is a provision stipulating that cooperative R&D has priority in selecting projects. Especially the cooperative format of R&D projects in the Industrial-base Technology Development Project is an important centre to be funded. In the Highly Advanced National (HAN) Project that was launched in 1992, the cooperative format is mandatory for project formulation (Table 3).

Since the format of joint research among industry, universities, and GRIs is preferred in evaluating and selecting the proposal of national R&D projects, the cooperation rate of national R&D projects is quite high. The overall cooperation rate in national projects is 58 per cent, which means that 611 out of a total 1048 projects are performed through cooperative research. The cooperation rates of individual projects are as follows: 87.2 per cent of the HAN Project, which amounts to W 75.5 billion; 46.1 per cent of the NSP, which amounts to W48 billion; and 85 per cent of Industrial-base Technology Development Projects, which amounts to W53 billion.

Table 3
Status of Cooperative R&D in National R&D Projects

Type of Project	*No. of Projects*	*Unilateral Projects*	*Cooperative Project*
Highly Advanced National Project	148 (100)	19 (12.8)	129 (87.2)

National R&D Project	727 (100)	392 (53.9)	335 (46.1)
Industrial-base Technology Project	173 (100)	26 (15.0)	147 (85.0)
Total	1048 (100)	437 (41.7)	611 (58.3)

Industry-initiated Cooperation

In the industrial sector, the number of cooperative research projects is listed in Table 4.

Table 4
Number of Projects and R&D Investment of Firms

Type of R&D	*No. of Projects*			*R&D Investment*		
	Large Firms	*S&M Firms*	*Total*	*Large Firms*	*S&M Firms*	*Total*
In-house R&D	2274 (117)	987 (150)	3261 (267)	281401 (110)	56669 (127)	338070 (237)
Joint R&D	158 (54)	151 (71)	309 (125)	34358 (49)	14882 (55)	49240 (104)
Commis sioned R&D	159 (42)	43 (27)	202 (69)	18959 (40)	1738 (24)	20697 (64)
Total	2591	1181	3772	334718	73289	408007

Note : The number in ( ) indicates the number of companies surveyed.

Source : Sung R. Cho, *et al*, Establishment of the linkages among Domestic R&D Institutions, STEPI, Dec. 1993.

In 1992, 284 private research institutes carried out 511 cases of cooperative R&D, which consisted of joint research and commissioned research. Out of the cooperative R&D projects, 309 cases (about 60 per cent) are joint R&D projects and 202 cases (about 40 per cent) are commissioned R&D projects. The proportion of cooperative R&D is 13.5 per cent (511 out of 3,772 projects, and the portion of joint R&D is amounted to W 69.9 billion (20.7 per cent of the total R&D investment of R&D private research centres). Among them joint R&D projects accounted for W 49.2 billion (14.6 per cent) and commissioned research projects, W 20.7 billion (6.1 per cent).

The R&D performance of industrial technology research consortia in 1992 shows that, out of a total 121 projects of 21 consortia, 81 projects (66.9 per cent) were cooperative R&D between consortium member companies or with other research organizations. Five projects (4.1 per cent) were commissioned projects while 17 projects (1496) were carried out by a single-member company and 18 projects (14.1 per cent) were performed by multi-member companies.

University-initiated cooperation

In the universities, as shown in Table 5, 972 projects were carried out by thirty university centres of excellence, science research centres (SRCs) and engineering research centres (ERCs) in 1992. Among them, the proportion of cooperative R&D between universities and industry was only 20 projects (2.1per cent). Compared with 1.5 per cent of 1991, the proportion increased.

Table 5
Forms of R&D Projects Performed by Centre of Excellence

Classification		*Total*	*GRI Commiss-ioned*	*Industry Commiss-ioned*	*Industry univ. coop-erative*
SRC (14)	91	278	246 (88.5)	27 (9.7)	5 (1.8)
	92	295	263 (89.2)	24 (8.1)	8 (2.1)
ERC (16)	91	526	409 (77.8)	110 (20.9)	7 (1.3)
	92	677	393 (58.1)	272 (40.1)	12 (1.8)
Total (30)	91	804	655 (81.5)	137 (17.0)	12 (1.5)
	92	972	656 (75.5)	296 (25.0)	20 (2.0)

Note: SRC — Science Research Centre; ERC — Engineering Research Centre.

Source: Korea Science and Engineering Foundation, 1993.

Most Popular Types of Cooperative R&D

In large companies the percentage of independent projects in 1992 was 87.8 per cent, while the percentages of joint research and commissioned research was 6.1per cent respectively. In small and medium-sized companies the percentages were 83.16 per cent, 12.8 per cent and 3.6 per cent respectively. The statistics tell that small and medium-sized companies relatively prefer joint research to other types of cooperative RE.

The most favoured partners in cooperative research initiated by industry are universities. The number of industrial cooperative R&D with universities was 319 cases in 1995, followed by cooperation with GRIs (245 cases) and with other companies (122 cases). Cooperative R&D with universities was usually in the form of short-term, small-amounted projects while cooperative R&D with GRIs was mostly in the form of long-term, large amounted projects.

The commercialization rate of industrial cooperative R&D projects with other companies was 25.2 per cent (29 out of 122 projects). The rate with consolidated reached 22.6 per cent (12 out of 47), the rate with GRIs was 11.8 per cent (24 out of 245), and the rate with universities, 154 per cent (44 out of 319). The reason that the commercialization rate of the cooperative R&D with GRIs was low is the cooperation between private companies and the GRIs is mainly carried out in the big science areas.

Supporting Systems for Cooperative R&D

An embryonic type of cooperative R&D first started in the 1970s when private firms commissioned R&D projects to GRIs. Nowadays, the government has adopted various promotion measures such as research grants and tax credits for cooperative R&D in order to encourage cooperative R&D activities.

First, the government enacted a series of laws that promoted technology development in Korean industry. The "Technology Development Promotion Law" was enacted in 1977. This law provided a legal basis for the establishment of an industrial technology research consortium. The "Industrial-Technology Research Consortium Promotion

Law", enacted in 1986, includes providing research funding and tax credits for research consortia[6]. A special fund, which was established under this law, guarantees collateral for loans that are used for technology development. This law also provides several tax benefits to research consortia. Research consortia that have more than 10 researchers are exempted from customs duty when the consortia import research equipment. The law also allows the consortia to delay income tax payment for a certain time period. The law supports research consortia through the government procurement mechanism under which products resulting from the commercialization process of research consortia are preferred in the case of government-procured equipment.

The "Science and Technology Promotion Law", enacted in 1991 encourages the frequent exchange of researchers and professors. The law guarantees the reclaimed status and compensations to exchanged scientists after they return to their original positions. Enacted in 1993, the "Industry/ Universities/Research Institutes Cooperative R&D Promotion Law" enables cooperative innovators to make it easier to obtain research funds, manpower, facilities, and the information necessary for cooperative research.

Second, the government expanded the coverage of tax benefits. The "Tax Exemption Law" (exempts 10-15 per cent of total R&D) investment from tax. Thus exemption is extended to the amounts paid to outside research institutes, universities, and research consortia when the firms commission R&D projects to them. The law applies not only to the costs incurred for purchasing technology from Korean nationals but also to the costs incurred for information collection. Third, besides the legal measures above mentioned, there are other promotional mechanisms. The government supports the Study Circle Programme that encourages holding meetings of researchers, engineers, professors, and businessmen away from their normal work-places and providing an operational budget to individual study circles. In 1994, there were 247 study circles registered to the Science and Technology Policy Institute. In 1994, the Consulting Center for Technology Application was established in Seoul and Taeduck together, which provides small and medium-sized companies with technical services. The services include information search for technical matters, and connecting services between firms and GRIs for technical services.

Success Factors and Difficulties

Successful Cases of Cooperative R&D

In bureaucrat-initiated cooperative R&D where the government and private firms matched funds for a large-scale project, the role of the GRIs was very essential for success. GRIs accumulated necessary technologies by using government funds transferred to them by participating firms in the consortium and helped private firms develop those transferred technologies, and the firms finally commercialized them.

Even though it is very hard to measure the performance of all cooperative RE activities, a few successful cases of bureaucrat-initiated cooperation R&D are outstanding. The development projects of electronic switching systems (TDX-1, TDX-10), memory chips (4-Mega D-RAM and 16-Mega D-RAM) and micro-computers (TICOM1 and TICOM 2) were successful cases that have utilized cooperative R&D format in the 1980s and the early 1990s. All of these developments resulted from the large-scale national R&D projects that were performed by consortia organized by the ETRI (Electronics and Telecommunications Research Institute) with the participation of four private firms. Especially, the development of commercialization of the TDX series was so successful that Korea became an exporting country of electronic switching systems instead of an importer. After the development of the TDX-1 through the consortium, the design technology was transferred to the participating firms for production. Later, the consortium continued in the second stage and developed more advanced switching systems called TDX–10.

Success Factors of Cooperative R&D

Success factors in cooperative large-scale national R&D projects were varied. The existence of concrete goals of cooperation R&D, the grant of incentives to the participants, superb roles of government as a moderator, close cooperation among participants and the existence of stable domestic markets for technologies can be listed as the possible success factors, but it is hard to distinguish the most important factors.

Despite the fact that there were several successful cases, it is arguable to maintain that overall evaluation about cooperative R&D in Korea so far is satisfactory. As mentioned above, it is true that most of the cooperative R&D activities was organized in the 1980s by bureaucrats. It would have been more desirable for individual R&D actors to organize and perform cooperative R&D by themselves rather than the government.

Difficulties in the Cooperative R&D in Korea

According to a survey about the difficulties faced by research institutes affiliated with private companies when engaged in cooperative R&D with other private firms, the disclosure of confidential technology is the most serious barrier to involving cooperative R&D. Other barriers include dispute over the distribution of research results, over-competition with competing companies, and the immature atmosphere for cooperative research. To overcome these barriers and to promote cooperative R&D it would be desirable to establish a special organization that selects cooperative R&D projects and plays the role of mediator between participants. In addition, national research projects should be research consortium as well as cooperative R&D among the industry, universities and GRIs. (Tables 6 & 7).

Table 6
Barriers and Solution for Cooperative R&D among Private Firms

Major Barriers and Solutions	*Degree*
Barriers	
• Revealing confidential part of technology	XXX
• Dispute over distribution of R&D outcome	XX
• Overcompetition among competitor companies	XX
• Immature atmosphere for cooperative R&D	X
• Technological gap between cooperating partners	X

Solutions	
• Establishing mediator organization	XX
• Expanding government's leadership in national R&D	XX
• Activating a research union	XX
• Supporting a R&D consortium on a regional basis	X

Note: XXX — Very high; XX — High; X - Average

The most serious barrier in cooperative R&D among the industrial, university, and GRIs is the lack of commercialization of R&D results. This barrier is due to the preference of university and GRIs for basic research, which is not directly related to commercialization demanded by industry. Another barrier is the shortage of information about the manpower and technology of counterpart organizations.

Table 7
Barriers and Solutions in Cooperative R&D among Industry, Universities, and GRIs

Major Barriers and Solutions	*Degree*
Barriers	
• Lack of commercialization	XXX
• Lack of information on manpower able to cooperate	XX
• Difficulty in checking partner organization's technological level	XX
• **Solutions**	XX
• Establishing information centres specialized in manpower and technology	X
• Promoting exchange of researchers	

Note: — XXX - Very high; XX — High; X — Average

Promising Signs for the Cooperative R&D in Korea

Inspite of the many barriers to cooperative R&D in Korea there are several promising signs indicating that cooperative R&D could flourish in the future.

Firstly, the "Industry-University-Research Institutes Cooperative R&D Promotion Law", enacted in December 1993, shows that the government has a strong intention to make cooperative R&D more active. It supports cooperative R&D activities by systematically rearranging the existing policies and institutions related to cooperative R&D.

Secondly, industry-supported science parks are being actively established in major Korean universities in order to implement cooperative R&D projects on a long-term basis by academia and industry together. For example, construction work for building the Techno-Complex at Korea University began in February, 1994. Won 54.8 billion will be invested to establish the Techno-Complex in which ten private firms will move in to participate in cooperative R&D. Other examples are the R&D institutes for semiconductor and new materials at Seoul National University; the Radio Wave-Communication Institute at YonSei University and the Korea Computer Graphics and CAD Consortium between SERI and KAIST. These institutes cooperatively perform R&D with participating private firms.

Thirdly, Industry-Academia or Academia-GRIs Cooperative Educating Programmes have been launched. The Industry-Academia Cooperative Educating Programme aims at developing S&T human resources. The programmes offer masters and doctoral programmes in the major universities for engineers and technicians working at firms. For example, KAIST now runs a programme that accepts part-time or full-time students sent by firms. The Academia-GRIs Education Programme is established for the purpose of providing university students with an opportunity to work at GRIs and accumulate on-site expertise. As of 1994, eleven GRIs, including KAIST and thirteen universities, including Korea University and Yonsei University, cooperatively run these programmes, which have an enrolment of 703 students.

Finally, the Study Circle Programme is successfully under way. The programme promotes exchanges of S&T information among researchers in various technological fields, which is a starting point for cooperative R&D. As of October, 1994, there were 247 study circles in various fields.

Conclusion

Korea's industrial development is assessed to be successful. One of the main reasons for Korea's remarkable achievement in developing industrial technology is that Korea was able to rapidly absorb and assimilate the necessary technological capabilities from abroad. The country, which heavily depends on imported technologies, now faces the problem of weak international competitiveness. It is getting harder to obtain the necessary technologies from foreign sources because of the recent technological protectionism enforced by technologically-advanced nations. Therefore, an analysis of the national systems of innovation is needed and measures to maximize efficiency and effectiveness of the systems are required. One of those measures is the industry-university-GRIs cooperative R&D.

Cooperative R&D in Korea has not been fully developed due to the unbalanced distribution of R&D resources to the three R&D actors in addition to heavy dependence on foreign technology. To respond to rapid changes in the world S&T environment, the government is actively pushing forward cooperative R&D as a way to utilize the technology potentials of the R&D actors. Even though some promotional policies and supportive measures of cooperative R&D have proven to be quite effective, an even more comprehensive and systematic approach has been required.

Responding to this request, the "Industry- University - Research Institute Cooperative R&D Promotion Law" was enacted. This law has made a great impact on the cooperation practices among the industry, universities and GRIs. However, problems still exist with the newly enacted law. One of the problems is how to make the law more effective. It is still difficult to coordinate policies and regulations planned and executed by various ministries. Cooperative R&D requires mutual trust and collaboration of all parties concerned. However, on account of the prejudice of our society in putting more value on teaching jobs, it is difficult to expect professors to readily move to industry.

Despite all these problems, cooperative R&D among the industry, universities, and GRIs has brought about considerable achievements. The enactment of the "Cooperative R&D

Promotion Law" is expected to function as a good starting point for systematically coordinating related regulations, for activating the exchange of manpower and technology information, and for building mutual trust among participants. Through these efforts, cooperative R&D activities will be improved in terms of quality as well as quantity.

REFERENCES

Bozeman, B., "Evaluating Government Technology Transfer" Can the New Cooperative Technology Development Paradigm Enhance U.S. Industries Competitiveness? International Conference on Changing Technology Environment and Policy Responses, Seoul, Korea, 1991.

Cho, M., A Study on the Classification of Forms of Industry, Academia, Research Institute Cooperative R&D, Korean Society for Policy Analysis and Evaluation, 1992.

Cho, S. R., *et al*, Establishment of the Linkages among Domestic R&D Institutions, STEPI, 1993.

Friar, J. and M. Horwitch, "The Emergence of Technology Strategy : A New Dimension of Strategic Management" Technol. Soci. 7, 1985.

Korea Industrial Technology Association, Major Indicators of Industrial Technology, 1993.

Ministry of Science and Technology, The Background of Enactment of Cooperative R&D Promotion Law and its Major Contents, 1993.

4

Institutionalization of Inter-Organizational Alliance*

N. Mrinalini

Cooperation for mutual benefits, among various institutions, both formally and informally, is not a new phenomenon. For the past two decades, one observes a prodigous growth in the formal collaborative arrangements of interaction, especially between the firms. Governments, through certain policy incentives and necessary infrastructure, have made efforts to forge cooperative participation of various participants in the economy. The reflections of these are the emergence of several cooperative arrangements like science/technology parks, consortia, cooperative research centres, etc. In the recent past, researchers from various disciplines like Economics, Sociology and Management Studies, have tried to focus on 'why two organisations (especially the firms) seek each other's cooperation in their activities'? The alliance is seen in terms of mutual benefits, like assets sharing, cost saving through resource pooling, sharing of the risk factor associated with R&D in the high risk areas, etc. Inter-firm alliances are broadly interpreted as a means to avoid market transactions.

Collaborative interactions can broadly be of two types: between two similar 'organisations' (inter-firm) and between

---

The author is extremely thankful to Pradosh Nath for sparing his valuable time for discussion, which helped in sharpening the focus of the study.

two disparate 'organisations' (firm and the government/ research institution). This process of cooperation can be visualized in the following ways:

1. Firms in a particular industry segment come together to undertake generic research where the cost, risk factor, gets distributed over the participating firms and at the same time each and every firm in this group can make specific use of the generic technology for its own purpose (like microelectronics).
2. The government initiates cooperation process through the creation of science/technology parks, cooperative research centres, etc. Long-term continuous interaction is expected to be facilitated because of the proximity of the various participants.
3. Establishment of certain short-term contractual collaborative arrangements between firms and research institution/university for specific tasks.
4. Government initiates the process of cooperation among various participants like research institutions, firms, universities for certain strategic goals to be achieved through joint efforts.

Organizational Arrangement for Collaborative Interaction

Any two organizations, whether similar or dissimilar, in their functions, commitments and orientation, have to shell out some bit of their autonomy, for entering into a collaborative alliance [Aiken, M. & Hage, J.,1968]. More so, for two disparate organizations, in which case, commitments, functions and orientations would drastically differ. The issue here is, how a cooperative alliance between two 'organizations' with different orientation and goals be forged. It would require reorientation in their management practices to incorporate accountability to their partner. So, if two organizations want to collaborate on a 'project', the presumption can be that the partners are expected to be benefited by this partnership to the extent that they are even prepared to give up a part of their autonomy. Mutual gains should weigh more than the losses for the alliance to come into existence. With this understanding, it is not very difficult to perceive the alliance between two similar organizations (like two firms) where the expectations from each other are of similar type. The cooperative alliances among disparate entities like government, university and industry cannot be a very spontaneous and natural process. In this case, there is a fundamental difference among the organizations with respect to their commitments, functions and orientations. For example, the industry's goal of increasing market share is very different from that of the university and research institutions, where the expected output is the number of research publications in reputed journals and the number of patents. Two divergent goals would mean organizational arrangements, having variance in their respective organizational management practices. In such situations, how can a collaborative alliance be forged between two dissimilar organizations? How is this process initiated? The study does not deal with the issue of why two different 'organizations' come together but rather, the focus is on how they are brought together, especially where the government has a mission and wants a cooperative alliance to be forged among the industry, university and the research institution.

How can the government really stimulate the process of interaction among various 'actors' ? In the case of government, the goal being some national mission, would require a long-term collaborative interaction among the partners. A closer interaction would enable better understanding of the need and also the capabilities to tackle the problem. The government has to take the initiative, as the benefit to the industry is not obvious to begin with. Confidence building in the heterogeneous partner (the firm) is a necessary step. Government's mission would require mobilization, utilization and creation of resources, for which a long-term interaction, as suggested by Aiken and Hage, a 'joint programme', is necessary [Aiken, M. & Hage, J., 1968]. A joint programme involves an organizational arrangement different from short-term work contracts. In essence, creation of new organizational arrangements, to trigger off a process for resource mobilization, creation and utilization, having broader economic and social implications.

There is not much literature available on the process of institutionalization of how the inter-organizational alliances are forged, especially when the collaborating partners are heterogeneous in nature. There are some references to inter-organizational arrangement with a focus more on the structure rather than on the process. Vinck, in his study, has dealt with a case of social importance, where the government took the initiative to mobilize the scientific community and to bring them together for a common cause [Vinck. D, 1993]. There is some elaboration on government's effort in mobilizing the concerned partners. Whereas, Apen *et al,* in their study have suggested the creation of a new organizational structure, a body, to forge partnership between the research institution and the industry [Apen,P.G. *et al,* 1994]. Liyange and Mitchell in their study have partly brought out the importance of having different organizational arrangements depending upon the orientation. The organizational structures of the cooperative research centres in Australia have different management practices like corporate, research and managerial, depending upon their orientation [Liyange, S.& Mitchell, H., 1993].

The focus of the study is to delineate the process of institutionalization of the process of cooperation, especially between a research organization and the industry, where the government has a mission. In this case it is more like an alliance between government and industry. How is this process of cooperation initiated? What is the role of the government in this process? These are some of the issues addressed in this study.

The process of institutionalization of cooperative alliance between government and industry is being discussed by taking cases from Taiwan and Japan. In both the cases, the government played an important role in bringing together the various players. Whereas in the case of Taiwan, the government had a mission for which the industry's participation was imminent, in the case of Japan, the mission of regional development programme was jointly evolved with the industry's active participation. So, in the case of Taiwan, the government had to initiate the process, as interventionist and facilitator, to bring the industry into the programme. In the case of Japan, as the industry was already a party to it, the government had to play the role of a facilitator. In both the cases the process of institutionalization of cooperative interaction between the government and the industry is different. In both the cases, the alliance between the research institutions and the firms (in this case the Small and Medium Enterprises (SMEs)) got evolved in the context of the strategic goal.

Government as Initiator and Facilitator

In Taiwan, SMEs form the backbone of their economy. With the international market becoming more and more technologically competitive, especially in the high technology areas, Taiwan government got initiated into a thought process to strategically develop an industry which would have an impact on other industries as well. In the seventies, the textile industry was the largest exporter and the second largest was the electronics industry. At that time the electronics industry in Taiwan was still an assembly industry and the government in consultation with the industry and the academia, decided to identify such a technology which would revolutionize the electronics industry and also would have a spillover effect on other associated industries [Chang, P. *et al.*, 1994]. This process resulted in choosing the IC (Integrated Circuit) technology and the semiconductor industry in the 70s to be the industry for development. This, being a highly technology intensive area, required heavy investment without having any surety of the outcome. Being a government's mission, initiative was to be from the government's side. The industry was neither in a position nor was willing to make any initial move in this direction. There were various steps through which this was institutionalized.

The first and the foremost thing for the government was to win the confidence of the industry. The government took the initiative in the form of setting up a task force funded by the Ministry of Economic Affairs, to work out the strategy for implementation. The industry and the government jointly made an assessment of the existing capabilities and infrastructure facilities in the area. As the SMEs in the information technology did not have the necessary resources to acquire technology and also to invest further in R&D for technological upgradation, the Taiwan Government set up two institutions namely Industrial Technology Research Institute (ITRI) in 1973 and the Institute for the Information Industry (III) in 1979, to extend full support to the SMEs. This was a very crucial step. As the industry was not technologically competent to make the right choice of technology for import and further development, the government, by establishing the research institute, took this responsibility. This was not enough, as the required manpower for furthering the activity in this area was not available in Taiwan at that time. The research institute also had to act as a mediator between the industry and the university. The important effort in this direction was a joint 'Multi Project Chip' programme, initiated by ITRI along with the National Science Council and the Ministry of Education, to generate the required manpower for this mission, which was non-existent at that time. To begin with 9 universities participated in the programme and ITRI through its computer-aided design training, encouraged the students and the professors from the universities to develop their own capabilities [Chang,P. *et al.* 1993a]. This initial effort triggered off interest in the academia to participate in large numbers resulting in the generation of the required manpower. Thus, the government provided the industry with the necessary manpower. This programme also ensured the active participation of the university resulting in bridging the gap of manpower requirement and supply.

The sourcing of technology was taken up by ITRI. Then the necessary R&D investments were made for improving the technology and also to upgrade its activities from technology improvement, to technology generation, and to new product development. Once the technology was developed, there was no taker from the industry and so the government had to prove the effectiveness of the technology. The Electronic Research and Service Organization (ERSO) was set up with the idea of facilitating the process of technology transfer. The task for ERSO was not only to develop the technology but also to set up a joint venture company with the help of ERSO's engineers and the capital provided by the government [Hou,C. & Gee, S., 1993].

In this case the government provided the necessary venture capital and also faced the risk associated with the initial launching of the new product and its acceptance. The research institute did this by setting up a spin-off firm, privately owned, having the advantage of the engineers who developed the technology. This is again another important initiative taken by the government, where the risk associated with the development and introduction of the new technology was borne by the government. Once the technology was tested, there were many takers from the industry, especially the SMEs, and this helped in attracting investments from private sector firms [Chang, P., 1993b]. By taking the initial risk associated with the launching of the new product, the government could succeed in diffusing the technology amongst the private firms. This was successfully followed up with the necessary infrastructure and supportive policy incentives to sustain the growth. In this case, the research organization has extended its service right from sourcing of technology to technology generation with supportive facilities, to transfer of technology, through joint ventures. The activities envisaged for the research organization were product development, generation of skilled manpower, to impart training to the firms on specific requirements and to provide incubation to new firms. With this sort of concerted goal-oriented approach where the government's role has been that of the initiator and supporter, facilitated the process of technology generation and transfer. This triggered off dynamism in the system resulting in the establishment of Taiwan's integrated circuit industry in the early 1980s. Taiwan's first IC manufacturing industry was United Microelectronics Corporation (UMC), a spin-off from ITRI.

There are several significant features in the development of Taiwan's semiconductor industry. There were various stages through which the government had to pass through to gain ultimately the confidence of the industry. The government at no stage could take the industry for granted. The process of institutionalization has been the government's direct involvement in the choice of technology, undertaking risk investment in R&D to build indigenous technological base and transfer of technology in the form of spin-off firms. This facilitated the establishment of a strong indigenous semiconductor industry in Taiwan. This process has helped in the creation of new organizational arrangements and new alliances which has introduced dynamism in the system. In addition to this, the government had several sector-specific incentives and provided the necessary infrastructure. It is like a total package where the government's role has been significant in initiating and stimulating the setting up of the semiconductor industry. It is more like a technology push, where the government had a definite role identified for the SMEs in the high technology area and the research organization's role was to provide the necessary technological inputs. In this case, the industry did not respond very spontaneously to the government's mission, but it was a gradual process of the government's effort to win the industry's confidence which resulted in an organizational arrangement involving all the three actors.

The other goal-oriented approach has been the regional development programme of Japan. Of late, regional governments, also known as the prefecture governments, have been emphasizing on regional development, by helping the local SMEs through prefecture government-owned, research organizations. The broad policy framework of prefecture governments is to evolve programmes for regional development. The regional development is thought of, in terms of development of small and medium enterprises of the region and also making use of local people, and local resources to a large extent for this purpose. It is a concrete programme where synergy amongst broad policy framework, local industry's needs and local skills is visualized. In this case it is a joint mission of the government and the industry. Unlike in Taiwan, here the industry is part of the decision-making process at the regional level.

Regional development programme is a vision which is jointly worked out by the government, academia and the industry. The research organization, the prefecture government and the local firms are involved right from the beginning and this mutual involvement helps in identifying the problem and hence, the transfer of the research results to the industry is not a problem. There is an active involvement of all the three actors.

So, the situation here is different from what one observes in the case of Taiwan. Here, the government's role is not that of an interventionist, but only of a facilitator. For decision-making purposes the government has not distanced itself from the industry but rather the industry is part of the system. The important issue in this case is the process of institutionalization of this dynamic interaction.

The prefecture governments have 'The Small Industry Promotion council', which has representation from local industry, prefecture, university and financial institutions, to work out the industry structure and trend forecast for five years [Ohnishi, Y]. From this programme a broad understanding of the requirements of the local firms is drawn out. Then, by taking up projects and action programmes, these problems are tackled. The research organization of the prefecture takes up projects based on these understandings.

In most of the cases, the establishment of the research organization has been the outcome of the joint efforts of the prefecture government and the local industries. In many cases the research organizations have been established based on the request of the local industry. The research organization's role is to cater to the needs of the industry and how it is to be done is decided at the prefecture level. The research organizations are directly under the department of industry, commerce and finance of the Prefecture government which again reflects the orientation of the research organizations.

In the already existing traditional areas, the research organizations help them to solve their problems, which could be in the form of consultancy, testing, standardization, tackling technical problems, etc. One such research organization is Iwate Industrial Research Institute. Though established in 1873, it got its present name in 1992, with a focus on technological upgradation of local traditional industries, particularly in ironware. At the same time new research organizations are created to undertake research work in the high-tech areas to provide the local firms with inputs in the high-tech areas. One such research organization is the Iwate Biotechnology Research Centre established in 1992 based on the decision taken by the Advisory Board of the senate of the prefecture and Board of Agriculture.

The mandate of these research organizations is to extend technological support to the small and medium enterprises of the region. The regional specificity gives them a well defined goal. All the programmes of research organizations are oriented towards achieving the ultimate goal of regional development through the development of the SMEs in the region. The recruitment of personnel to these research organizations is a centralized process. Preference is given to local people as they have better interaction with the local SMEs. The research organizations have to have closer interaction with the clients to have a better feel of their problems.

This is one system where the government which funds the research organization is also its client. The decisions are taken at the prefecture level regarding the utility of the research results. The local firms for whom the research organization is supposed to cater to has representation in different government forums. There is concurrence between the government policy framework and efforts undertaken by the government in achieving the same.

In this approach, the mandate of the research organization is in accordance with the broad policy framework of the regional prefecture government. The research organization is the means through which the objective of the prefecture government is achieved. The three actors, the research organization, the government and the local industry are actively involved in working out the programmes for achieving the goals.

Conclusion

Alliance for strategic goal triggered off the process of new alliances and new relationships. Government has facilitated the creation of a dynamic relationship among the participants. The dynamism is sustained through the process of institutionalization of new alliances, new relations in certain new organizational arrangements. The process of institutionalization of this cooperative alliance has been different in both the cases. In the case of Taiwan, where the strategic goal was to establish an industry in the high technology area, it set *pace* for the alliances, emergence of new organizations and new relationships. The government through its research institutions got the imported technology assimilated and upgraded before giving it to the firms. This helped in building the confidence of the industry. The initial risk of introduction of new technology was totally taken on by the research institute. The government also provided the venture capital for the spin-off firm to establish the credibility of the technology. This is to say that the research institute, when once it got convinced of the performance of its technology, started transferring it to the firms. Right from the choice of technology to its implementation it has been institutionalized through interactive participation among the industry, government and the research institute where university has also played an important role in the form of generating manpower as per the requirement of the industry and the research institute. The electronics industry could get established because of the concerted effort of all the three partners with government taking the initiative.

In the case of Japan, where the strategic goal was regional development, the research institute, the regional government and the firms interacted closely to articulate the need. Here, participative decison-making of all the participants concerned with regional development, is institutionalized wherein, accountability is built into the system. In this case, the industry was developed to the extent that it could participate actively in the government's decision making process. This helped in achieving the goal as the users of the technology were part of the system which decided upon the research activities.

From both the cases the important issue which emerges is that new relations and new alliances get created in the process of achieving certain goals and what is important is to institutionalize these goals in the form of organizational arrangement.

REFERENCES

Aiken, M. & Hage, J. (1968): Organizational Interdependence and Intra-Organizational Structure; *Ameri. Sociolog. Rev.*, 33, 1912-930.

Apen, P.G. *et al* (1994): A New Model for Public-Private Partnership; *Technol. Soci.*, 16(4), 2389-402.

Chang, P. *et al.* (1993a): Taiwan's Approach to Technological Change: The Case of Integrated Circuit Design, *Technol. Anal. Strat. Mgmt.*, 5(2), 3173-178.

Chang, Pao-Long, *et al.* (1993b): Linking Technology Development to Commercial Application, *Int. J. Technol. Mgmt.*, 8 (6/7/8), 4697-712.

Chang, P. *et al.* (1994): The Formation Process of Taiwan's IC Industry-Method of Technology Transfer, *Technovation*, 14(3), 5161-171.

Hou, C. & Gee, S. (1993): National System Supporting Technical Advance in Industry: The case of Taiwan, in Nelson, R.R. (ed): National Innovation System, A Comparative Analysis, Oxford University Press, New York, 6384-413.

Liyange, S. & Mitchell, H (1993): Organisational Management in Australian Cooperative Research Centre, *Technol. Anal. Strat. Mgmt.*, 5(1), 73-14.

Ohnishi, Y : Role and Strategies of Public Research Institutes for Small and Medium Enterprises in Japan, Asian Productivity Organization, Tokyo, (mimeo).

Kojima (1994): Technology Support System for Japan's Casting Industry, Department of Liberal Arts, (mimeo).

Vinck, D. *et al.* (1993): A Network Approach to Studying Re search Programme: Mobilizing and Coordinating Public Response to HIV/AIDS' *Technol. Analy. Strate. Mgmt.*, 5(1), 839-54.

5

Human Resource Development: A New Role for Advanced Research Institutions

S. Visalakshi

The new developments in Science & Technology (S&T) and international politics leading to globalization and opening up of economies have led to an international competitiveness. The concerns of improving productive performance led to an emphasis on technological advantage as a means to substantive gains in these. An increased use of S&T personnel is required for devising, developing and introducing new technologies which can fire the engine of growth (Collins, 1988).

An adequate supply of qualified personnel is a precondition to development based on a strongly knowledge-intensive economic growth. According to a report by Organization for Economic Co-operative and Development (OECD) (1989), a shortfall is envisaged between the supply and demand for R&D personnel or qualified knowledge-workers in the exact and natural sciences in the OECD countries. The studies by Yamada (1987) for Japan, Collins (1988) for the USA, Pearson (1988) for the UK and Dekkar *et al.* (1994) Berendsen *et al.* (1995) for the Netherlands on the employment situations for the R&D personnel have also come out with similar concerns. Though the studies have concentrated on the gap in R&D personnel demand and supply, they include both scientific and technical skills. These studies have served as inputs into the policy for human resource development in these countries. Some of the policy options suggested involve the following:

- make the position of R&D work more attractive by different means (short-term strategy);
- promote education in the exact sciences (long-term);
- increase capital-intensive research (substitution of labour by equipment);
- invest in organisational improvements;
- increase investments in training and improve quality of available human capital;
- retain S&T personnel in science and engineering work;
- recapture and retrain those originally trained in science and engineering but now out of field;
- continue absorption of personnel prior to completing formal degrees.

While there is a concern the world over about the mismatch of the R&D and S&T personnel supply and demand in the years to come, due to a variety of reasons like:

- ageing,
- rapid development of certain technologies which result in obsolescence of the knowledge of researchers, and
- decline in college age population,

in India, we face another mismatch, i.e., in addition to the above in the numbers, in the skills and knowledge between the supply and demand. While the numbers is a concern of the institutions which generate human resource in the country, the advanced R&D organizations can take part responsibility in abating the crisis and help relieve the concern regarding mismatch in the skills of educated S&T manpower. In this study an attempt is being to made to outline the reasons for the occurrence of mismatch in skills in supply and demand of human resource in S&T in India and policies and actions which may bridge the gap between these. It further delineates, in this context, the possible role of national R&D institutions and suggests mechanisms for achieving this, quoting some of the already existing examples, mainly in the development of Biotechnology (BT) in India.

Background

In India human resource generation has been the State's

responsibility falling under State's subject according to the Constitution of India. Though there is an indication of a mandate to maintain parity in standards and compatibility there have been lots of problems. There had not been well-drawn out long-term and short-term plans of HRD in this country. Establishment of a university, college or school (technical) was often based on political reasons. There has been no concerted effort to assess the requirements in numbers, levels and types of skills matching the requirements of the clients. Even if there were a few sporadic attempts they have not resulted in a manpower generation system which is more responsive to the needs. Before getting into the reasons for the observed mismatch in demand and supply of skills, it is worthwhile to dwell on the mechanism (model) of HRD and utilization in India.

Resource and Control	*Generators*	*Users*
S T A T E	Schools (Technical)	R&D organizations
		Indian industries
	Colleges	Institutions/Industries abroad
	University	Bureaucracy Policy-making bodies

The model does not have any feedback. The generating system is isolated and insensitive to the changes in the needs of the clients. Due to lack of interaction, the generating system continues with offering courses based on traditional discipline-oriented patterns. The organizational structure and culture of functioning in colleges and universities are department-centred and have very little scope for inter-disciplinary courses. The university system is infra-structurally not geared up sufficently to take up training in modern fields.

The industry in India which thrived basically on internationally transferred, tried and tested (should we say also, abandoned?) technologies wherein the skill requirements were very limited, did not come out in the past with specific demands of skills. As far as the number of personnel required is concerned it was to an extent met.

There were no criteria of efficiency and excellence attached and accountability demanded of the HR generating system supported by public funds, except in terms of numbers rather than in terms of absorption rates of personnel who turn out of the system. The other clients like research organizations could reorient the available personnel to match their requirements due to a more facilitating, flexible and better equipped organizational structure and functioning.

Due to the above reasons in the past the HRD in India could go on with generating and absorbing systems isolated from each other, moving their own ways, mostly public funded as a charity without sufficient sensitivity to the needs (and changing technology regime). The needs that existed were more for production which too was based on standardized proven technologies from abroad on turnkey basis. This gave a limited scope to look for great skills at invention and innovation but only for standardization and trouble-shooting. The industries had lower levels of ambition born out of less or no competition in the protected market constituted by buyers of very low level awareness of product quality and price. Investments in R&D or new and high- tech skills were not compelled and hence investment capability, which was quite low was used for buying (repeatedly) technologies (even if they were old) from abroad.

In the case of the research organizations also there were no serious compulsions of accountability in terms of time and objective achievements in government-funded institutions. This allowed intake of less matched or non-matched skilled manpower and to gradually train in newer specializations and newer techniques/skills. The luxury of ample time and well developed infrastructure allowed the existing modes of HR generation in this country.

Though there are other organizations where natural science and professional courses graduates were absorbed like in the government departments and institutions abroad, for the present study the emphasis would be restricted to research organizations and Indian industries.

Coming to the current scene, there are changes which

has made the three major actors in the HRD who could afford to stand at a distance making unkind remarks about each other, to come closer and recognize the positive points in each other in the system, break the ice and talk about their problems and try to devise systems and methods to achieve common benefit to all the three. The major factors which brought the three actors closer have been:

- new developments in science and technology leading to new technologies with certain characteristics (e.g. biotechnology (BT) & information technology).
- new economic policy at the national level (opening of the economy for global operators).
- developments at the international level (globalization, GATT, TRIPS).

Due to the compulsions arising out of the changes the three participants involved in the HRD are made to see a common benefit in producing manpower in matching numbers with matching skills. The objective to generate manpower compatible with local industries would attract students and funds from industries which would increase greatly the sustainability of the generating component of the system. This also reduces the quantum of investments by individual firms/companies.

The developments due to new technologies which are science and information intensive are:

1. Shift from the production skill dominated requirement pattern of manpower by industries to high science based skills,
2. Need for inter-disciplinary skills and dependence of R&D intensive product manufacture,
3. Increased investment towards upgradation of the existing skills on a continued basis,
4. High rates of obsolescence of technology leading to intake of manpower with new skills/knowledge.

The changes which have come up due to the new economic policy are:

- the availability of better quality products manufactured elsewhere or by MNCs in India.
- increase in the awareness of consumers about quality and existence of better alternatives.

- need to compete at quality and pricing to cater to local market by the Indian industry.
- increasing awareness among the industry of the role technologies, which are information/high science intensive, play in the value addition of products [Ghosh, 1996].
- limitations in the industry to investment in technologies with very short life cycle.
- need to invest in management and marketing skill development.

The implications of the changes at international level leading to globalization of markets and regulations :

- stringent quality control (Q.C.) requirements (ISO 9000, etc.) and standards requirement of different countries;
- competitive pricing requirement;
- to achieve the above two requirement of sufficient quantity and quality skills at R&D and Q.C., etc.;
- new skills in packaging and marketing;
- patenting requirements, etc.

Other developments in the recent years at the national level like gradual but steady withdrawal of funds by the State to education and R&D have created new responsibilities for organizations involved in these to attract funds for survival and success which demands higher levels of accountability. This increase in the level of accountability disallows mainly two things, namely insensitivity of the education system to its clients' needs and the lack of concern for time-budgeting and quality service at R&D institutions. The above developments have led to a need to generate manpower compatible with the demands which is continually changing on an immediate basis.

Due to the fact that the nature of demand is continuously changing and involves new skills, all the three participants in the HR generation and utilization, namely educational, R&D institutions and industry require to work in conjunction. They need to understand their,—

- interdependence,
- strengths,
- weakness.

and devise mechanisms to make use of the complementary capabilities in each other. Thus the model suggested involves—

- cooperation
- interaction
- and mutual complementation.

This model assumes sharing of inputs and responsibilities by all the partners leading to a common benefit of better survival/performance and benefit to the society in which these actors are situated and functioning leading to development which has positive implications for all the three. The scheme could be best represented by Fig. 1.

Some experiences of advanced and newly industrializing countries in the area of biotechnology, which is one of new technologies based on high science and intensive information inputs, could illustrate the point made.

Observations on HRD in Biotechnology in Advanced Countries

The two predominant features of manpower needs in biotechnology are their multidisciplinary nature and high qualification profiles. According to general consensus, such multidisciplinary training in biotechnology should complement rather than supplant the training of specialists within the underpinning disciplines. The ultimate goal of such training should be to prepare individuals for successful interaction in research teams. In addition, the exposure to practical industrial problems as a part of the training process has also become recognized as an efficient means of facilitating rapid integration of R&D manpower into industrial biotechnology.

It is the expectation of some analysts that BT would generate a large number of new jobs but these would require very specific highly trained individuals. Some of the earlier studies have shown that this fact has been perceived by the respondents of the survey from high-tech firms who rank quality education, and availability of skilled workforce as the factors most crucial for successful commercialization of BT in the USA.

Attempting to overcome the obstacles, which early

educational traditions raise like the separation of biological sciences from the engineering faculties, most of the advanced countries, wherein development of biotechnology for industrial use has occurred, have responded to the demands of biotechnology by reorienting traditional training programmes in order to achieve desired interdisciplinarity. These efforts include institution of special interdisciplinary bachelor's degree programmes and short courses at the post-graduate level for academics and industrialists. Many countries strove to increase interaction between the educational institutions and industry at the training level in biotechnology. In this section some of such attempts and strategies adopted are outlined.

The United States of America

The currently working biotechnologists are not trained in BT but are drawn from related fields like Biochemistry, Molecular Biology, Genetics, Microbiology, Botany, Plant Pathology, Virology, Biochemical Engineering, Fermentation Technology, Immunology, etc.

Different studies done earlier show that though there is an agreement among different forecasting groups about the fact that there will be an increase in need for personnel in BT in the nineties, there is no agreement about the types of jobs which might be available and the type of training required for these jobs. Office of Techology Assessment (OTA) (1988) estimates are that the personnel who would be employed in companies involved in BT based product commercialization which includes both dedicated BT companies (DBCs) and large diversified companies, would be about 36,000 of whom slightly more than a half would be scientists and engineers. (This is a five-fold increase over the situation in 1983). According to a report on a survey of California firms using new BT by Blumenthal *et al.* (1987) the professional (scientific) and technical personnel occupy 63 per cent of total manpower, while clerical workers (17 per cent), managers (15 per cent) and floor level production and maintenance workers would be the rest of the total BT labour force (5 per cent).

Understanding the importance of the highly qualified and skilled personnel, in order to attract such skills the American BT companies have allowed freedom to their scientists in both pace and direction of their work, and accorded flexibility of work environment akin to the academic ones. The available literature on the status of manpower involved in BT in different countries [OTA, 1984], clearly draws the picture of strengths and weaknesses of that country in their ability to produce BT based products and their competitiveness.

Table 1
Major Categories of Biotechnology R&D Personnel in Firms in the United States (OTA/NAS Survey)

Area of technical expertise	*Present employees Number*	*per cent of total*	*Employees to be hired in the next 18 months Number*	*per cent of total*
Areas related to genetic manipulation				
mol. genetics	586	23	302	25
Hybridoma antibodies	247	10	146	12
Plant molecular biology	76	3	63	5
Areas related to scale-up/downstream processing				
Microbilogy	334	13	160	13
Biochemistry	326	13	125	10
Bioprocess engineering	186	7	100	8
Areas related to all aspects of biotechnology				
Enzymology immobllized systems	219	9	59	5
Cell culture	187	7	66	5

Source: OTA, 1987.

Tables 1 and 1A give indications of the categories of personnel who are involved in BT related work in the US BT firms.

- Molecular biologists 26 per cent (only 3 per cent are plant molecular biologists),
- immunologists 10 per cent i.e.,

About one-third of the technical personnel employed in 95 of US BT companies covered by Office of Technology Assessment (OTA), and National Association for Science (NAS), survey are specialists in genetic manipulation with training in molecular Biology and Immunology.

Specialists in scale-up and downstream processing: Specialists in scaling up and production of genetically manipulated microbes and other organisms, (GMOs) and purification of products are,

- Bioprocess engineers — needed for designing, constructing and maintaining scale-up equipment and also in bioprocessing;
- Biochemists involved in recovery, purification and quality control (QC) of protein products;
- Shortages were felt by the companies in the supply of biochemists;
- Pharmacologists and toxicologists;
- Hybridoma and cell fusion technique experts;
- Bioprocess engineers;
- Industrial microbiologists.

[Shortage has occurred due to the fact, that the Federal support was more for basic research.]

Table 1A

Shortages in Major Categories of PhD biotechnology R&D Personnel in Firms in the USA (OTA/ NAS Survey)

	Numbers of Firms		
Area of technical expertise	*Experiencing shortages and planning to hire in next 18 months*	*Experiencing shortages and not planning to hire in 18months*	*Not experiencing shortage but planning to hire in next 18 months*
Bioprocess engineering	11	1	15
Recombinant DNA	10	1	29
Gene synthesis	7	3	7
Plant molecular biology	4	4	15
Industrial Microbiology	3	4	14

Source: Office of Technology Assessment, 1987.

To cope up with the requirements of manpower in new and emerging areas of biotechnology the attempts made in the USA includes taking up new initiatives in biotechnology. Table 2 gives an indication of the number of such new initiatives taken in the 1980s by the American colleges and universities to meet the education and research needs of the upcoming biotech industry. According to a survey by OTA, about 60 such initiatives have been taken by 49 different US colleges and universities. The identified initiatives in training in biotechnology include programmes ranging from 2-year applied associate of science degrees to short courses in particular biotechnologies designed for professional scientists.

The other types of initiatives are situated outside academic institutions in the research centres of biotechnology (university-sponsored). These generally do not sponsor courses but lend their support to biotechnology education and training through access to equipment, faculty development and provision of research training/opportunities for both graduates and undergraduates. They also serve as a focal point for discussions of the best means to educate and train new biotechnologists.

Specialists in genetic manipulation:

Table 2
University Initiatives in Biotechnology Training

Year	*No. of Initiatives*
1980	2
1981	2
1982	5
1983	9
1984	8
1985	6
1986	9
1987	13

Source: Office of Technology Assessment, 1988.

Another interesting and important fact relates to the involvement of industries in the development of these programmes. According to the OTA survey of 1988, out of the 41 programmes of the above-mentioned nature, 34 had been evolved in consultation with industry. Consultations with the industry included surveying local biotechnology companies and sending programme proposals to the industry representatives for comments. But such coordination among the developers of different programmes was not observed. Though the response of the industry to the personnel turning out of such programmes could not be observed, the personnel coming out of these new initiatives faced no difficulties in finding employment in their fields of training.

Japan

In the early 80s Japan experienced shortages of experts in

- genetic manipulation;
- hybridoma work;

due to less emphasis given to basic biological sciences in the universities. Strategies which were taken up to overcome the shortages were,—

- in-house training by Japanese companies;
- hiring Japanese researchers from abroad;
- sending employees abroad for mid-career training;
- migration of researchers from other Japanese companies (despite Japanese feature of lack of personnel mobility).

Strengths;

- bioprocess engineers;
- industrial microbiologists;
- experts for scale-up.

This led to a possible domination of Japan in the international amino acid market. This was achieved through their chemical industries using biological process and also their university system which kept a close contact with the local industry and oriented itself to lead to out turn of applied biochemists and process engineers.

Federal Republic of Germany

Weakness:

rDNA and hybridoma but taken care of by making it a priority area at the university education level.

Strength : Scale-up and bioprocess engineering.

UK

Adequate strengths in rDNA and hybridoma specialists. Strength in bioprocess engineering and scale-up capabilities. Serious concern for the shortages which occurred due to braindrain of BT experts to other countries. There is no information available to us regarding the strategy or efforts taken by the UK to overcome this.

Switzerland

Taking care of the shortages perceived in skills called for getting workforce for BT from other countries by attracting them through high wages and better working conditions or by locating R&D abroad. Academic institutions show strength by giving emphasis to both basic and applied research relevant to BT.

France

Situation in the 1980s

Weakness —	Bioprocess engineering Microbiology Molecular biology Enzymology Hybridoma experts Plant and cell culture experts

Pockets of excellence in:

- Immunology - (Pasteur Institute)
- Enzymology (Tech. Univ. of Compiegne)
- Bioprocess engineering (Tech. Univ. of Compiegne) are noticed.

Shortage felt at all levels, namely:

- scientists,
- engineers,
- technicians,

State-initiated programmes in the form of mobilization programme (1982-86) and national programme (1986-90) to give an impetus to the development of BT. One of the major objectives of these programmes is to develop competence in areas like microbiology, enzymology, fermentation, genetic engineering, improved seeds, plants, vaccines, logistics and supplies of equipment and generation of R&D (skilled) manpower. To the above programmes, the public research institutions responded positively: (a) A number of universities brought the course work and research in fields supporting BT to state- of- the-art levels, (b) Interaction between research institutions and industrial community increased greatly (6 per cent of all industry-university; scholarships were in BT by 1991) (Ramani, 1996).

Apart from the above-mentioned country-related shortages in general, the deficiency in skills related to the following specializations are identified by different studies like:

- Protein chemistry (Bernstein, 1987; Rosazza 1987)
- X-ray crystallography (Gwynne, 1987)
- Cell culture (Humphrey, 1987; Hyer, 1987)
- Quality control and scale-up related specialists (Covel, 1987; Wang, 1987)
- Microbial ecologists (Tauf, 1987; OTA, 1988).

According to Vetter, 1986, sufficient supply is noted in the case of specialists in, —

- Molecular biology,
- Biochemistry,
- Cell biology,
- Immunology.

It is envisaged that there could be shortages in newly emerging fields like protein engineering for the following reasons :

1. Difficulty of predicting manpower demands in a new field,
2. The time lag required for educational institutions to prepare for new fields,
3. Expense of new faculty and newer equipment prevent institutions to shift rapidly to new areas,
4. Shortages in the labour force at the level of training in a new field (Gwynne, 1987; Makulowich, 1987).

Microbial ecology which otherwise was an obscure speciality, of late, has become important in the context of release of GMOs and increasing awareness, regulations of Environmental Protection Agency (EPA), and an increased demand causing shortages is expected to occur in the near future.

There is an expectation of increase in demand overweighing the supply for bioprocess engineers which has not become true due to the low-volume, high-value nature of current BT products. This might change when many products with larger volume requirement reach BT markets. Still it is not expected to exceed 10-15 per cent of the total BT labour force.

The need for immunologists is predicted to increase beyond supply capacity due to increased commercialization of Monoclonal Antibody (MAB) based products and AIDS research.

Some strategies to cope with the shortages :

1. Mobility among disciplines, e.g., molecular biology researchers took care of shortages in plant molecular biologists.
2. Shortages for biochemical engineers was taken care of by traditional chemical engineers.
3. Response of universities with increased emphasis in the areas where shortfall is envisaged.
4. Efforts made to automate part of the production process to reduce the demand. (This may not be an adaptable strategy in the Indian situation).

While these seem to be of consequence for a short-term, the American experience seems to hold the key for long-term strategy to take care of reducing the mismatch between the demand and supply of skills required by BT industry. The strategy involves partnership of the industry and academic institutions with a certain amount of input from the research organizations. The nature of involvement could be varying while the benefit is shared. It is also observed from the existing literature that the USA has been the most successful country in the commercialization of BT and sufficient manpower to achieve this, indicating the validity of partnership between all the three actors in the system involved in the HRD. The other interesting feature is the noticeable absence of the State in this effort.

It can also be noticed from the survey of the efforts by different countries other than the USA that in most of the countries there had been an interaction between the BT industry and university system to abate crisis of shortages in the skills experienced/envisaged by the industry.

With this background if one sees what sort of developments occurred in India in the case of HRD in biotechnology, one would be able to see the differences in strategy and efforts and results out of it.

Indian Efforts in HRD for Development of BT

Keeping in view the inherent quality of BT R&D and production namely:

- Very high rates of obsolescence;
- R&D-intensive nature;
- Skill dependence;
- Many parts of procedures not being amenable to automation;
- Incompatibility to standardization of many processes;
- Need for constant monitoring of structures (modified, introduced by BT methods), etc.

The need for developing infrastructure in terms of skilled manpower was considered very important by the Department of Biotechnology (DBT). DBT was created in the year 1986 out of National Biotechnology Board (NBTB) formed in 1982. DBT initiated programmes for a planned development of Biotechnology in India. The efforts included long-term and short-term plans. The Department concentrated on two areas, giving emphasis to manpower development and creation and strengthening of infrastructural facilities. The objective of these efforts is to build up trained manpower in biotechnology which has the competence required for development of viable BT industry in the country.

As a result of efforts on the creation and strengthening of infrastructural facilities for research and teaching in the form of equipment making available research reagents, around 50 institutions in the country came into being, who are in a position to teach and also research in BT. This is further strengthened by project grants by DBT for more than 100 projects (till 1990) and organized centrally coordinated projects. Some of the facilities which support the R&D in BT that are set up by the DBT efforts include :

- germplasm banks for animals, plants and microbes;
- animal houses housing those required for preclinical trials;

- facilities for maintenance of cell lines of animals and plants;
- facilities for oligonucleotide synthesis;
- making available the reagents, radio-labelled compounds and enzymes required for BT research;
- upscaling facilities in the form of pilot plants;
- facilities for protein and peptide sequencing;
- national NMR facility;
- antibiotic development consortium.

Coming to the other thrust area of manpower development, DBT initiated programmes for postgraduate and postdoctoral teaching in biotechnology. This was achieved by,—

- creating faculty positions for M.Sc./M. Tech. and post-doctoral training programmes in about 25 chosen universities and other institutions;
- attracting students for these by providing stipends;
- a placement cell was created for helping the students after training to guide in getting suitable positions.

Apart from these long-term training programmes, several short-term courses ranging from 10 to 16 every year have been conducted from 1984. The main purpose of this programme is to help life scientists in the university and research institutions or in-house R&D units of the relevant industries to reorient themselves to meet the immediate requirements. Around 1100 people were trained under this programme. In addition to these, the department has initiated short-term training programmes for technicians in the industrial R&D as a mid-career training programme. To cope with the demands of BT industry for technicians a programme of one-year diploma courses for science graduates has been started in a modest way.

To further strengthen the competence of the already existing R&D workforce in BT, both national and overseas associate ships are awarded annually for advanced research and specialized training in the areas identified as very important. So far, around 280 such associates have availed themselves the scheme which is expected to fill the gaps in capabilities and demand for highly skilled manpower in the country.

Mechanisms to Achieve New Model of HRD in India

Mechanisms of achieving HRD which do not suffer infirmities of mismatch of skills and at the same time utilize the capabilities/strengths inherent in the three sectors could include a new role for the advanced R&D institutions.

The R&D institutions have evolved over the years to have, —

- well-trained manpower in different specializations in science, technology, medicine and agriculture;
- the infrastructure development in some of Indian institutions is comparable with their counterparts elsewhere in advanced countries.
- there are capabilities and infrastructure available in the R&D organizations to impart training in new technologies like biotechnology, information, aerospace technology, etc.

Table 3 gives an idea of the number of formal training in BT conducted by different agencies including advanced R&D institutions in the country. They also are involved in training manpower in BT at the doctoral level. Almost 90 per cent of R&D institutions are involved in this activity in various fields, both traditional and modern interdisciplinary fields.

Table 3
Distribution of Training Providing Organizations in BT (1983-1992)

Nature of Offering Institute	*No. of Res. Training courses*
University	18
CSIR	5
ICMR	2
ICAR	3
DBT	4
DAE	1
DST	1
Others	5

Table 3A
Participants in Short-Term Training Courses in BT

Year	*Universities*	*CSIR*	*Res.Councils & Others*	*Industry*
1983	17	2	5	0
1984	14	1	8	1
1985	38	13	18	14
1986	114	15	37	10
1987	95	21	32	4
1988	46	7	20	3
1989	34	8	40	0
1990	45	8	39	2
1991	69	9	39	19
1992	23	2	17	0

Development of biotechnology in India can serve as an illustrative case for a cooperative mode of HRD of matching skills of demand and supply where an important new role is played by the government and R&D organization. The case also highlights the reduction in the rates of success by the lesser involvement of industry. A recent NISTADS study points out the persisting gaps in the demand and supply of skills for industries involved in BT related production in the pharmaceutical industries in India (Visalakshi *et al.*, 1995), is one of the consequences of industry's behaviour. Table 3A on the pattern of participation in the short-term training courses organized by DBT in biotech from 1984 to 1994 shows the hesitation of the industry to participate in these training courses aimed at skill upgradation for BT R&D and production related functions.

In the case discussed, we see an example of how mutual complementation could generate more suitable BT manpower using the resources and strengths of the three participants involved in a mission. As stated earlier, still there are gaps and envisaged gaps between required skills and available skills by the end of the century in certain specializations as per a study by NISTADS (Visalakshi and Sharma, 1995).

The above observations of developments in different countries in the manpower development in biotechnology make it amply clear that a successful commercialization of new technologies requires manpower development only when there is an involvement of industry, R&D and educational institutions at equal level participating in both stakes and benefits. When the industry is replaced by an agency like the State as in the case of India, the results do not match those observed in the case of the USA, Japan or Germany (See Tables 4 & 4A).

Table 4
Distribution of Firms Active in Biotechnology by Country

Country	*1969-84*	*1985*	*1990*
USA	349	495	>700
JAPAN	112	304	>450
UK	34	148	~250
WEST GERMANY	51	59	N.A.

Source: Adapted from Luigi Orsenigo (1989) for data up to 1985. Data for 1990 collected from various sources by the author.

Table 4A
Country-wise Distribution of Patents and Publications

Country	*Publications*	*Patents*
USA	183	199
JAPAN	60	232
GERMANY	32	63
FRANCE	33	21
UK	25	19
USSR	26	15
OTHERS	17	

Some of the suggestions to reduce these gaps required as per our proposed model:

Involvement of Industry

- as a resource provider;
- designing courses of interest in the universities and colleges (e.g. SPIC co-sponsoring B.Tech. in biotechnology course in Anna University);
- in monitoring the course of interest to them;

- initiation of collaborative research with academic or R&D institutions which would enable training of their personnel during the project.

The industries could either singly or through consortium approach provide resources for training of manpower they require at various levels, from lower end technology based skills like those required for plant tissue culture companies using micropropagation technique or personnel required for performing diagnostic tests at pathological laboratories, etc. This would also help in designing the courses with change in their requirements which come in an evolving industry like the BT industry.

R&D institutions should continue with their present involvement in providing,—

- formal short-term (BT training programmes), medium-term programmes (e.g. those organized by Central Leather Research Institute, Madras, CECRI, Karaikudi) providing courses in specialized skills;
- training postgraduate students for their doctoral and post-doctoral work in the fields in traditional and modern disciplines needing interdisciplinary inputs; Should also involve themselves in, —
- participating in university curriculum development
- partake in teaching of graduate and postgraduate level students in new and emerging areas;
- extend infrastructural facilities for students from universities which are poor in these;
- take up collaborative projects with university departments which would upgrade the skills of the university faculty.

Universities on their part would, —

- make themselves open and responsive;
- involve industries who would like to be benefactors by providing funds, create faculty positions and other facilities for starting new courses and in the curriculum development along with the involvement by the experts in the R&D institution [US example as in BT (OTA, 1989)],
- incorporate in the curricula information on the industrial work culture, in-house training in the

industrial units exposing the personnel to the problems both technical and managerial and troubleshooting methods which both university faculty and the S&T personnel at the R&D institutions are not much conversant with. This would increase the involvement of industry personnel in teaching in academic institutions and the interaction thus enhanced would lead to a better appreciation of each other's capabilities and limitations and development of trust which is a precondition to cooperation.

Role of State

The State should play the role of conductor of an orchestra or director of play or a coach of a team by,—

- initiating interaction;
- laying down the rules for such efforts;
- facilitating such efforts through a congenial policy environment;
- establishing a reward system which takes cognizance of such efforts and stay in the background.

Conclusion

Changes in nature of emerging technologies, economic scene at the national level and trade and other functions at international level have generated situations which necessitate State, research institutes and industry assume new roles in the generation of human resources, in Indian context. These participants are compelled to work in partnership in the generation of (manpower) human resources which matches with the requirement of industry. In the proposed model, the State's roles could be restricted to only as an initiator and facilitator of this cooperation wherein the participants engaged in the partnership of generating human resource of high value complement each other based on their relative strengths. Both share responsibilities, cost as well as benefits by this mutual complementarity. This is supported by the experiences of different advanced countries in the area of biotechnology which has all the characteristics of a new technology.

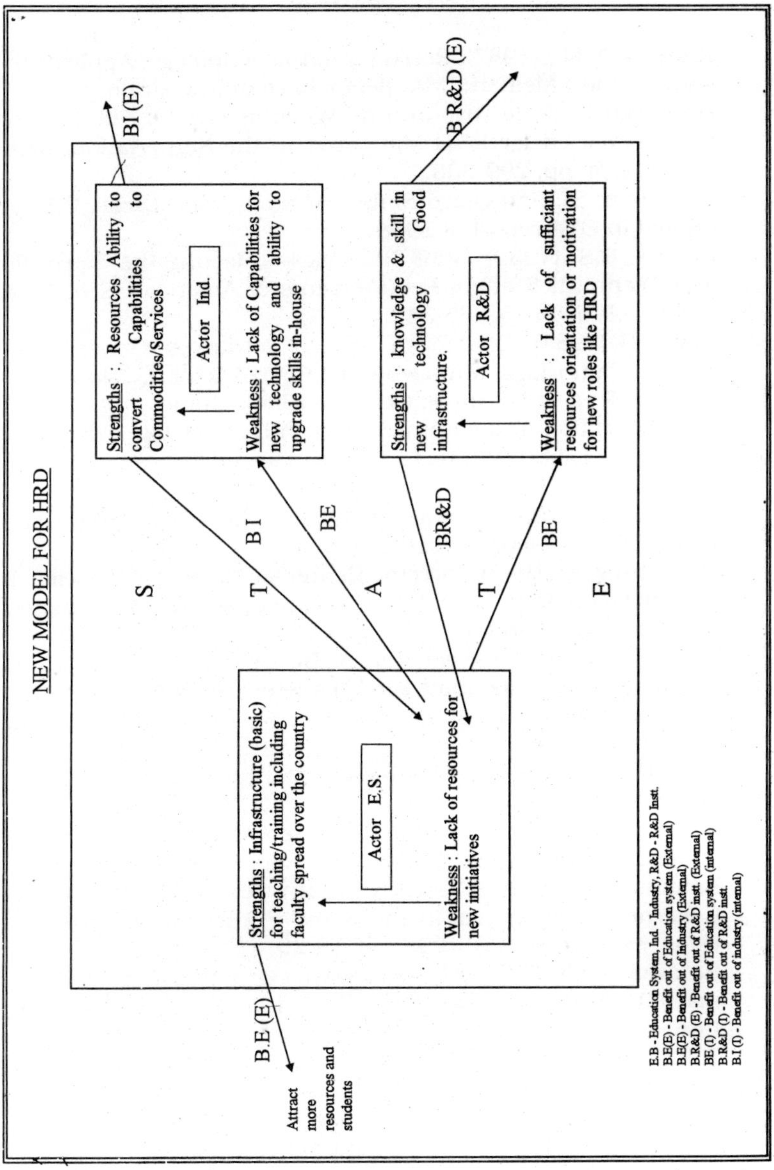
NEW MODEL FOR HRD
Strengths : Resources Ability to convert Capabilities to Commodities/Services
Actor Ind.
Weakness : Lack of Capabilities for new technology and ability to upgrade skills in-house
BI (E)
Strengths : knowledge & skill in new technology Good infrastructure.
Actor R&D
Weakness : Lack of sufficiant resources orientation or motivation for new roles like HRD
B R&D (E)
S
T
A
T
E
B I
BE
BR&D
BE
Strengths : Infrastructure (basic) for teaching/training including faculty spread over the country
Actor E.S.
Weakness : Lack of resources for new initiatives
B.E (E)
Attract more resources and students
E.B - Education System, Ind. - Industry, R&D - R&D Inst.
B.E(E) - Benefit out of Education system (External)
B.E(E) - Benefit out of Industry (External)
B.R&D (E) - Benefit out of R&D inst. (External)
BE (I) - Benefit out of Education system (internal)
B.R&D (I) - Benefit out of R&D inst.
B.I (I) - Benefit out of industry (internal)

71-A

REFERENCES

Andelman, M., (1987): Recruiter in Biotechnology, Applied Resources Ind., Medford, MA, personal communication.

Brendsen H, A de Grip and de Williams (1995): The Future Labour Market for R&D Manpower in the Netherlands, *R&D Mgmt.*, (25), pp. 299-308.

Bernstein, S., San Diego State University, San Diego, CA, as quoted in OTA report of 1988.

Collins, E.S. (1988): Staff Shortages— Meeting the Scientific and Technical Staffing Requirements of American Economy, *Public Policy*, pp.15, 335-342.

Covel, B. (1987): Director of Human Resources, Genetics Institute, Cambridge, MA, as quoted in OTA Report, 1988.

Dekkar, R. A. de Grip, Heijke, H. (1994): "Indicating the Future Education in the Netherlands" in Hans Heijke (ed), Forecasting the Labour Market by Occupation and Education (Kluver Academic Publishers, Boston), pp. 55-83.

Gwynne, P., (1987): X-ray Crystallographers Wooed by Drug Firms, *New Scientist* ,1(11): 5.

Humphrey, A. (1987): Director, Center for Molecular Bioscience and Biotechnology, Leigh University, Bethlehem, PA, as quoted in OTA report, 1988.

Hyer, Q.C.(1987): Association of Biotechnology Companies, Washington, DC, as quoted in OTA Report, 1988.

Kincannon, K., Kincannon and Reed, St. Louis, MO, personal communication.

OECD (1989): Research Manpower: Managing Supply and Demand, Paris.

Pearson, R. (1988): Scientific Research Manpower: 'A Review of Supply and Demand Trends' IMS Report No.169, Brighton.

Ramani, S.V., (1996): Creative Incentives: A Comparison of Government Strategies in India and France; *Biotechnol. Develop. Monitor.* No. 26, 1996, pp. 18-20.

Rosazza, J.P.N., Biocatalysts Research Group, College of Pharmacy, University of Iowa, Iowa City, IA, personal communication.

Taub, F., School of Fisheries, University of Washington, Seattle, WA, personal communication.

Congress, Office of Technology Assessment (1984): Commercial Biotechnology: An International Analysis, OTA-BA-218 (Elmsford, NY: Pergamon Press Inc.).

U.S.Congress Office of Technology Assessment (1988): New Developments in Biotechnology: US Investments in Biotechnology, OTA-BA-360 (Washington, DC, U.S. Government Printing Office).

Visalakshi. S., Sandhya G D, & Abrol. D.(1995): Assessment of R&D and production capabilities in the Pharmaceutical Industry in India in the context of BT Commercialization, NISTADS Rep.No. REP-156/95.

Visalakshi. S. & Sharma L (1995): Manpower Projections in Biotechnology in India by 2000 A.D., NISTADS Rep.No. Rep-159/95.

Yamada K (1987): A Simulation study on future demand and supply of researchers, Research Institute for Higher Education, Hiroshima, Univ. Hiroshima.

6

Role of R&D Institutions in Human Resource Development: A Case of Indian Paper Industry

N.J. Rao

The Indian economic and industrial scene has witnessed a major change since 1992 with adoption of the new economic policy. The impact of these changes has a liberalized economy with an open-door policy. This has opened up a new order of competition on a global basis. The protection provided to the industry is steadily being withdrawn. Further, as a signatory to the GATT agreement in 1994, India will have to honour all the principles in it. This will mean Indian industry will face severe competition in domestic market which will emerge from three sources. Imported goods come into India at low tariffs. Goods will be produced in domestic market by foreign owned/controlled enterprises using their brand names and technology. Goods will be produced in domestic market by nationally owned enterprises. As a result, competition will intensify.

How will domestic manufactures fare in such a fierce competition? What level playing field can be provided to Indian industry to face the liberalized imports and foreign direct investment?

The main factors which will cause difficulty for Indian industry to face the international competition include, —

- the size of our operations is small;

- the technology of our operation is low/obsolete;
- cost of inputs is out of tune with international prices;
- specific consumption of electricity, utilities and raw materials is higher;
- infrastructure costs are higher across the board;
- financial costs are out of tune.

These factors will force the Indian industry to look at options which will provide it with a competitive edge. Modernization will be required with associated technological upgradation to increase productivity, reduce costs, improve quality, improve energy and environment management efficiency, and improve financial management. All these will call for a qualitative improvement in human resource used by industry. There is a need to focus on developing a competent, forward looking human resource which can convert the challenges into opportunities and make the Indian industry truly internationally competitive and sustainable. Can R&D institutions play a positive role in this? We will try to look at this picture with the example of the Indian paper industry.

Indian Paper Industry — Status

The Indian paper industry has made rapid strides since 1812 when the first factory was established at Saharanpur. From 45,000 tonnes per year in 1930, India's installed capacity today is over 3.954 million TPA in paper and board sector and 0.315 million TPA in newsprint sector (Table1). The industry has a very wide base with plants of varying size and age using a variety of raw materials and producing a wide variety of papers and boards. The industry constitutes one of the important sectors of India's industrial economy.

The demand for paper and board has been growing. The challenges confronting the industry are many. These include fibrous raw material shortages, obsolete technology, energy shortages, and greater environmental consciousness. This is now compounded by the impact of economic changes under liberalized industrial policy.

In order to meet these challenges there will be greater stress towards,—

- Increased automation, instrumentation, computerization, and microprocessor based controls;

- Increased productivity;
- Increased machine speeds;
- Improved product quality at reduced costs;
- More exports. Joint ventures in India and abroad with tie-ups;
- Trading by international companies in India and through India.

Table 1
Paper Industry in India: 1992-95

	1992-93	*1993-94*	*1994-95*
General economic Growth rate (%)	4.3	4.3	5.8
Industrial growth Rate (%)	2.3	4.1	8.0
Paper and board: Installed capacity (Million tons/year)	3.55	3.79	3.954
No. of mills	340	350	380
Production (million tons/year)	2.143	2.329	2.510
Growth (%)	9.6	8.6	7.2
Imports (000' TPA)	39.16	46.82	69.73
Newsprint: Production (million tons/year)	0.312	0.361	0.402
Imports (million TPA)	0.243	0.311	0.292

Modernization and expansion of mills with improvement in technology level will be attempted to meet the increased market demand on the one hand and to achieve greater efficiency in energy and environmental management on the other. Medium and small agrobased mills will seriously attempt at putting up chemical recovery units to process black liquor. Greater use of waste paper for quality end products will be attempted. Mills will strive to reduce water consumption through water system's closure.

The human resource in India is one of the finest and it

can become a major strike weapon in globalization of paper industry. Indian manpower will find its own level in terms of wages and job opportunities in international paper market. The impact of this will be felt in Indian industry in terms of greater mobility of manpower. Shortages in competent manpower capable of handling new technology will be felt.

Performance of The Paper Industry

The average unit size in Indian paper industry is small. The level of technology is low. The raw material base is weak. The units are essentially of multi-machine and multi-product type.

The data on operational parameters in terms of raw materials, water, chemicals (caustic soda, salt cake, chlorine, alum, rosin, lime), coal, steam and power are analyzed. The figures show a wide variation among mills.

The specific energy consumption, water consumption are higher in Indian mills than international level primarily due to the types of raw material being used and the small size of the Indian mills (Table 2).

Table 2
Performance Norms of Indian Paper Industry
Operational Parameters

Detail	*Units*	*Large Integrated Forest-based*	*Medium Agro-based*
Water,	m^3	161 - 400 (277.8)	90 - 393 (208.5)
Caustic Soda	kg/t	13.4 - 131 (55.5)	1.24 - 250 (187.9)
Salt Cake	kg/t	20 -72 (42.7)	-
Coal	t/t	1.3 - 4.2	0.77 - 2.0
Chlorine	kg/t	20 - 156 (68.9)	25 - 262 (145.5)
Steam,	t/t	6.4 - 20.5 (10.13)	3.3 - 7 (5.77)
Power,	kWh/t	1193 - 2311 (1579.1)	841 - 2648 (1286.5)
Generated	% of total power	26.6 - 71.4 (47.4)	9.1 - 667 -

(*Note* : Figures in parentheses are the averages.)

(1992-93) Survey of 10 large integrated mills 13 small/medium agro-based. Total capacity, 0.92 million tons.

Regular Emps.	61.8 - 81.7	58.9 - 94.7
% of Total Emp.	(72.6)	(82.2)

Economic parameters like output/G.F.A. (gross fixed assets), sales turnover/G.F.A., capital productivity, capacity utilization and raw material storage are analyzed. The variations are wide.

Table 3
Economic Parameters

Detail	*Units*	*Large Integrated Forest-based*	*Small/Medium Agro-based*
Capacity Utilization	%	73 - 132.9	56.7 - 137.4
Output/GFA,	TPA/ Million Rs.	29 - 154.6 (51.6)	131 - 174.2 (80.8)
Sales/GFA,	Rs./Rs.	0.49 - 2.99 (1.08)	0.65 - 2.02 (1.57)
GFA/ Employee	Millions/ Total Employees.	0.119 - 0.736 (0.394)	0.184 - 0.645 (0.243)
Output/ Employee	TPA/Total Employees	11.3 - 33.2 (19.2)	7.39 - 35.76 (18.69)
Sales/ Employee	Millions Per Yr./ Total/Emp.	0.197 - 0.611 (0.373)	0.275 - 0.488 (0.389)
No.of Emp./ 1000 TPA Prod.	Total Emp./ 1000 TPA	30.2 - 88.5 (52.1)	26.9 - 135.4 (53.5)
Other Parameters			
Raw Material Storage	Equivalent Months of	0.87 - 5.33 (2.19)	1.8 - 13.74 (5.56)
Digesters	No./100 TPD	(3.55)	(18.3)
Paper M/cs.	No./100 TPD	(2.03)	(4.53)

GFA: Gross Fixed Assets (1992-93)

Though the capital intensity is larger in integrated mills compared to small agro-residue mills, the sale per unit gross fixed assets is of similar order. The capacity utilization shows a wide variation among mills and is generally higher in integrated mills. Inventory of raw materials per unit production is high in agro-residue mills (Table 3). Labour/manpower indices are evaluated. Labour productivity and sales per employee are nearly of the same order in large and small agro mills. The number of employees per unit production is of the same order in larger and smaller mills. The ratio of contractual to regular employee is 0.377:1 in large integrated mills against 0.216:1 in small agro mills (Table 3).

The manpower utilization is analyzed on qualification-wise distribution, levelwise distribution, occupational patternwise distribution and departmentwise distribution. 79.5 per cent of total manpower in large mills is engaged in production, services, maintenance and R & D functions while the figure is 80.2 per cent in small agro mills. The skilled/unskilled manpower component in small agro mills is 77.9 per cent, while the corresponding figure in large integrated mills is 87.22 per cent. Manpower with technical qualifications is 10.61per cent and 9.13 per cent of total manpower in small agro and large integrated mills respectively. Persons with high school or below qualifications are over 76 per cent in both category mills. Engineering degree holders, diploma holders, ITI/skilled workers are 2.27 per cent and 2.46 per cent, 3.04 per cent and 2.63 per cent and 3.82 per cent and 5.53 per cent respectively in large integrated and small agro mills. The percentage of persons with formal qualifications in pulp and paper/chemical engineering background is very small (Tables 4-8).

Table 4
Occupation Patternwise Distribution of Manpower (per cent)

Detail	*Large Integrated*	*Small Agro*
Professional, Technical and Related	18.81	22.64
Administrative Executive	6.22	5.13
Service Workers	21.27	9.97
Production and Related Workers	44.79	
Clerical and Related	8.91	14.78
TOTAL	100.00	100.00

Table 5
Levelwise Distribution of Manpower (per cent)

Detail	*Large Integrated*			*Small Agro*		
	Higher	*Middle*	*Skilled/ Unskilled*	*Higher*	*Middle*	*Skilled/ Unskilled*
Production	0.54	3.29	-	0.68	4.49	-
R&D	0.10	0.36	-	0.13	1.50	-
Other Tech.* Functions	0.75	4.09	-	1.08	5.04	-
Non-Tech.#	0.42	3.22	-	2.01	7.15	-
TOTAL	1.80	10.96	87.22	3.90	18.20	77.90

*Includes materials and stores, maintenance, utilities and services, civil works.

#Includes sales, finance, accounts, administration and personnel.

Table 6
Sectoral (Departmentwise) Distribution of Manpower (per cent)

Sector	*Large Integrated*	*Small Agro*
Production	35.57	54.35
R & D	0.76	
Marketing and Purchase		
Finance	9.79	11.07
Corporate Office		
Others*	43.94	25.83
Forest and Raw Mat.	10.02	8.76
TOTAL	100.00	100.00

*Large proportion is in service/utilize and maintenance functions.

Table 7
Qualificationwise Distribution of Manpower (per cent)

	Large Integrated		*Small Agro*	
Technical		9.13		10.61
Degree	2.28		2.46	
Diploma	3.04		2.63	
ITI	1.67		2.56	
Skilled	2.15		2.97	
Science		4.86		4.72
Arts/Commerce/ Others	5.81		6.59	
Management		1.40		1.26
High School and Below		78.81		76.81
High School	28.19		30.35	
Below High School	50.62		46.46	
TOTAL	100.00		100.00	

Table 8
Breakup of Technical Manpower (per cent)

Specialization	*Large Integrated*			*Small Agro*		
	Degree	*Diploma*	*ITI+ Skilled*	*Degree*	*Diploma*	*ITI+ Skilled*
Pulp and Paper	0.229	0.623	0.111	0.56	0.54	0.25
Chem. Engg.	0.423	0.022	0.0	0.29	0.01	
Mech. Engg.	0.913	1.371	2.124	0.70	1.11	3.22
Elect. and Inst.	0.50	0.649	1.486	0.43	0.67	1.99
Others	0.221	0.367	0.963	0.43	0.23	0.16
TOTAL	2.276	3.039	3.818	2.46	2.63	5.53

HRD Institutions Catering Exclusively to Pulp and Paper Industry

The Human Resource Development (HRD) activities in paper industry are to be taken up by several agencies. These include:

1. Teaching and training institutions (universities/ academic schools/engineering-technical institutions) associated directly or indirectly with pulp and paper activities like; IPT, LIT, etc.

The educational and training facilities in the area of pulp and paper technology are analyzed in terms of types of courses, seats available, staff position, infrastructure facilities, industry interaction, job opportunities for students and budgets. The facilities for HRD are inadequate. The institutions have poor budgetary provisions, inadequate infrastructure, weak staff structure and poor industry interaction.

2. R&D institutions in public and private sectors associated with pulp and paper and related activities like CPPRI, FRI&C, etc.

The R&D institutes attach considerable importance to forging of close links with the relevant sectors of the industry. They endeavour to make available to the industry the services of their staff and special facilities available in their laboratories and workshops for the purpose of dealing with the problems posed to them. The aim is to channelize the available expertise and facilities for application to design, development and experimentation in industrial problems, processes and projects. The most important aspect should be relevance of the activities of R&D centres to the immense problems faced by the country and the people. A large fraction of our intellectual community resides in centres far removed from where the action lies. R&D work of institutes must have industry relevance.

There is little interaction between the industry on the one hand and research and academic institutions on the other. The major reasons appear to be,—

- anxiety among academicians and researchers about secrecy in industry arising out of proprietary interests;
- industry's view of research and academy personnel to be in "ivory tower", isolated from the real world;
- the feeling among research/academy personnel that solving industry problems does not enhance their reputation or career;
- industry regards academic and research outputs as a mass of disjointed publications of little or no real use;
- market forces do not encourage the development of indigenous solutions to the problems of industry; and
- industry finds it convenient to go abroad for solutions to their problems as well as for new technologies.

3. Consultancy organization/manpower development institutions associated with conducting short-term courses in pulp and paper related topics. There is very little activity at this level now in the area of human resource development.

4. Paper industry's (through their personnel/HRD departments in mills) efforts for training fresh recruits and for training their employees periodically to increase skills/knowledge levels. There is some activity in some forward-looking mills. These programmes are general management oriented or technical in nature.

5. Industry associations like IPMA, IAPMA, AISPMA, PMMA, IPPTA, FICCI, etc. involved with pulp and paper industry. There is very little HRD activity except for conducting seminars/workshops.

Currently except for training/teaching of personnel by academic institutions before entry into paper mills very little

HRD activity is undertaken by other agencies listed above. The interaction among one another is also weak. With economic liberalization and opening of the markets, the competition is likely to become more intense requiring the industry to be more efficient and cost-effective. This would require upgrading of mills and the quality of manpower in terms of skills, knowledge and ability to perform. The trends will be to have tie-up for upgradation of technology and human resource. This can be achieved through greater interaction among different stake holders to prepare specific tailor-made training modules for target-oriented audience.

There is need for collaborative action. We should establish pathways which will permit engineers and managers in industry to visit and work in R&D centres for functionally meaningful periods of time to receive continuing education and to get their mental batteries charged. Visiting faculty appointments offered to senior personnel in industry will help to impart to our research and study programmes the flavour of industrial viewpoint and trigger research projects of immediate relevance. Where bridges of cooperation have been built, we should nurture them carefully and remove road blocks, if any, that hinder the movement of people. Where such bridges do not exist we should build them to make possible a network of goodwill, understanding and a coordinated effort between R&D institutions and industry to be based on a subtle relationship in which the autonomy of R&D institutions is utilized and complementary roles of R&D institution and industrial enterprise is allowed to fructify. Technology and its development are immensely important intellectual endeavours that demand the very best of human creative skill.

Challenges Before Human Resource Development Activity

The HRD utilization indicates the following trends :

- Growing tendency to utilize contractual/casual employees for activities like raw material handling and procurement, sludge and solid waste handling, sanitation, housekeeping, transport, watch and ward, etc.;
- Independent management of maintenance and ser-

vices functions will give way to integrated production and maintenance functions;

- Annual Maintenance Contracts (AMC) being given to specialist agencies for specialized equipment growth;
- Increased activities in R&D and in HRD to meet the challenges of the modernization, rebuilding, automation and quality demands will occur. A cultural change is taking place where old technologies and plant practices are giving way to modern high-tech processes and processes;
- There will be greater efforts towards productivity improvements with increased efforts to reduce manpower utilization. Energy and environment cells will be formed in most units;
- Raw material generation/procurement and marketing (internal/international) areas will receive more focused attention.

The existing HRD institutions have served the industry to some degree. Their effectiveness has been limited due to very little involvement of industry, poor budgetary support, inadequate infrastructure facilities and limitations on improvement in teaching standards and methods.

The R&D institutes have the problem of converting suitable technology to adaptation stage, a current weak link. Further R&D institutes are required to catalyse these activities through suitable brainstorming session, an area where inertia seems to exist.

The manpower assessment for the future has to be made keeping the following parameters in mind :

- requirements for new green-field units;
- requirements as replacements for the retiring components in existing mills;
- improvements in skills and continuing education of existing personnel at various levels;
- meeting requirements of manpower for globalization activities;
- meeting manpower needs of auxiliary sectors (R&D, design, consultancy, fabrication, wires, felts, special chemicals, etc).

The industry will be required to involve itself more closely with HRD activities at unit level as well as at HRD centres in terms of greater commitment, financial provision and by making available faculty and infrastructure. There has to be some degree of accountability of HRD activities.

HRD institutions will be required to reorient the programmes to meet the challenges of the industry's needs. Curriculum changes, bringing new inputs, faculty and laboratory upgradation, transfer of shopfloor information to classrooms, conduct of relevant short-term courses and training workshops are some areas needing attention. The industry interaction have to be increased manifold.

Future Human Resource Needs Assessment

The personnel needs for industry will be for the following areas/levels :-

- production and service related functions;
- HRD, design, troubleshooting functions;
- commercial activities including marketing, raw material procurement, finance, etc.;
- executive/managerial functions including corporate office;
- shopfloor operators/workers (production and maintenance);
- supervisory personnel (production, services, commercial functions);
- managerial/executive personnel/professionals (production, commercial function, R&D, HRD technical function).

Need will be for a competent, well-trained human resource to provide the task force for current activities as well as to act as thinktanks for tomorrow.

There will be a redistribution of the manpower with changes in needs, in levels of education and training. The component of qualified and unskilled workforce will diminish if not vanish. The HRD will be utilized as the major strength for cost reduction, quality improvement and competitive edge. Accordingly, the future manpower needs are shown in Table 9.

Table 9
Future Manpower Usage

Capacity Range	*Persons/1000 TPA*	
	New Green-Field Unit	*Existing Units*
300 - above	8-10	10-15
150 - 300	10-12	12-18
30 - 150 TPD	15-20	18-25
Below 30 TPD	20-30	25-40

Model for Management of HRD for Paper Industry

The proposed model for an HRD programme for the paper industry must basically meet the following objectives :

- to bring about a rationality in the structure, curriculum, duration, quality of the programme being run by various HRD institutions for first entry levels. The programmes have to be need based (at universities/teaching institutions);
- Monitor the institutions/programmes in terms of funding, infrastructure, industry interactions;
- Help in planning/coordinating service personnel programmes (continuing education) of various durations on topics of specific interest. The identification of topics, preparing of training packages and trainers are going to be major tasks of R&D institutions;
- R&D institutions must act as pilots in preparing relevant short-term programmes on new emerging technologies of relevance for the industry. This will act as a stimulant to the industry in reorienting itself to adapt and successfully operate new processes/technologies;

These activities should be through a "National Advisory Board for HRD in Paper Industry" formed at national level by Development Council, Ministry of Industrial Development, AICTE and with active involvement of HRD institutes/industry and professional bodies like FICCI, IMPA, IAPMA, IPPTA — (See Figure).

The conduct of HRD programmes at first entry level must be left to teaching institutions which should be helped and monitored for quality inputs. The existing R&D centre and one new centre in another part of the country can be made into a 'Nodal HRD Centre' or 'Technology Centre' for course formulation, monitoring and upgradation of resource persons/teaching materials. These HRD centres must be adequately funded to improve infrastructure, improve industry interaction, involve plans of hiring industry personnel (serving/retired) for class-room inputs, move class-room teachers to industry shopfloors.

New programmes have to be planned for ITI/worker levels with skill inputs. This is going to be quite challenging as the facilities in the area are practically non-existent at the moment.

The development of HRD programme for in-service personnel will pose maximum challenge (See Table10). This activity can be done broadly at two locations, namely; mill site or in HRD institutions. Demand-driven, economically viable, self-sustaining systems for operators, supervisors, line managers and other personnel of paper industry should be the main objective. This can be achieved by developing three or four Regional Training Centres (RTC) with the involvement of a core plant and nodal centres.

The RTCs will coordinate with mill, core plants, nodal centres (technology centre) involving suitable strategies in conduct of programmes. The National Advisory Board (NAB) along with nodal centres should formulate strategies for identification of programmes, preparing training materials (manuals, visuals, video texts, case studies), training the trainer and providing experts systems. The RTCs will concentrate on worker level/operator level mill based programmes for skill improvement while HRD centre should concentrate on programmes at other levels. This model can be broadened after initial experiences.

Table 10
Continuing Education for Existing Mills and for New Capacity

Level	*Details of Training*
Worker/Operatior	Skill development in operational/ maintenance functions at plant level. Tasks identification of training modules, training modules preparation, trainer development. Most difficult due to large number involved, language/communication, trainer shortages.
Supervisory	Knowledge and skill development at plant/ institutional level operational/maintenance functions with — quality/cost/productivity emphasis. Tasks-identification of training modules, preparation of training packages, development of trainers.
Middle/Junior Manager/Engineer	Knowledge development with some exposure to skills. Inputs relating to equipment, design, technology, commercial aspects. Programmes at plant/institutional level. Tasks-identification and preparation of training packages. Need high communication level.
Non Technical Managers/ Executives	Knowledge development in Pulp and Paper Technology and Correlation of non technicals aspects to manufacturing and productivity. Short duration programmes at metro politan centres/HRD institutions/mills/corporate office. Need high communication level.
Senior Level	Short duration state of art/knowledge expos rue for stimulating ideas. These brain storming sessions can be conducted with high order of communication skills in corporate offices/metropolitan cities on specific subjects.

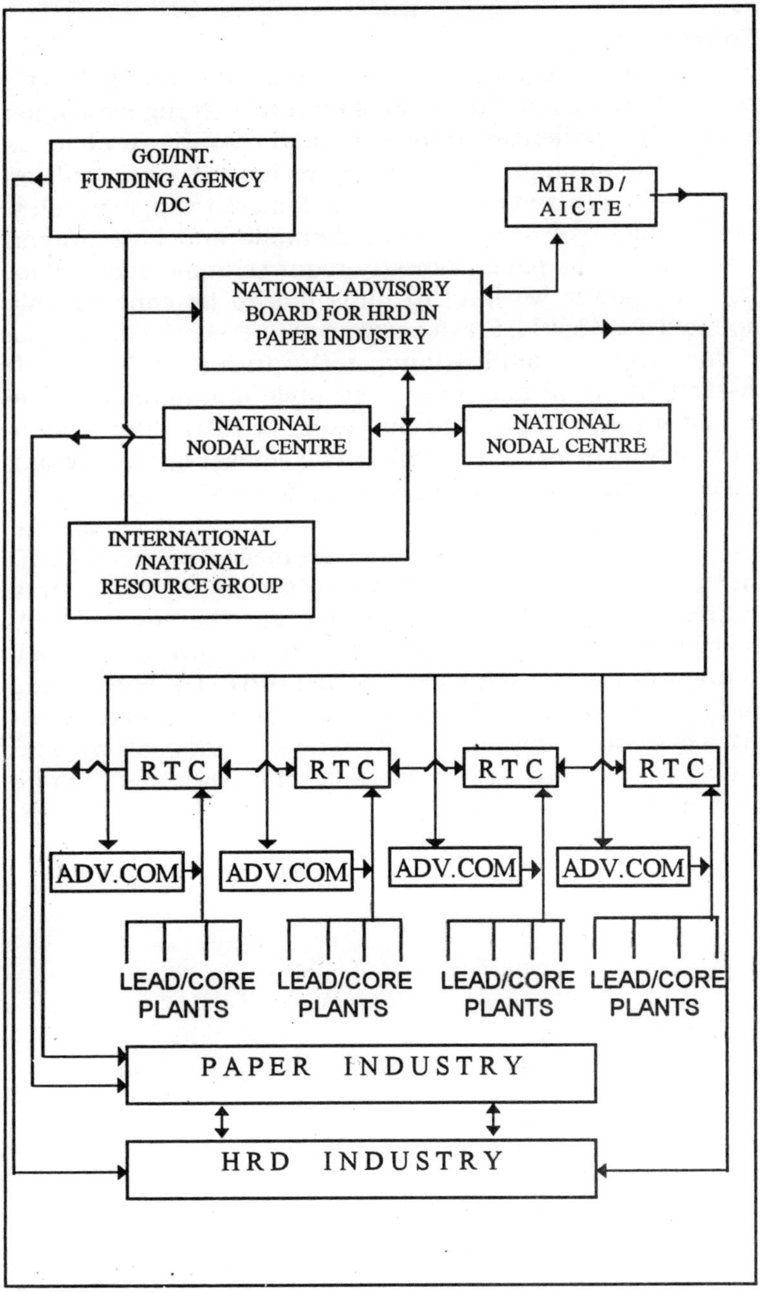

Fig. 1. Proposed Model for Human Resource.

Conclusion

The Indian paper industry is at crossroads. The challenges before it are many. The options require a strong motivation to face the challenges. Proper human resource at all levels will be the strongest foundations to launch an assault on the challenges and convert the challenges to opportunities. Development of a competent, dynamic and able human resource for the paper industry requires immediate action. The manpower working in mills has to be continuously updated at all levels. Fresh entry personnel should be capable of meeting the aspirations of the industry. The HRD institutions have to change their style of functioning. The industry must respond to the needs of HRD activity more positively. The need of the hour is a concerted effort with a direction. The time available is short. Results can be achieved if complacency gives way to action, dormancy to commitment. The captains of industry, the governmental agencies/R&D institutions and HRD centres must sit together for a brainstorming and success will come. The current half-hearted approaches to HRD activity must give way to well thought-out, purposeful and targeted HRD. The path is long and arduous. The challenge is great but with commitment the scene will change and the Indian paper industry will find itself as an Asian Tiger and the human resource will be second to none in the world.

7

Industry-TI Partnership in India: Select TI Case Studies and Implications*

Shekhar Chaudhuri

In recent years there has been an increasing concern regarding the effectiveness and efficiency of technology institutions. At an international seminar in early 1995 the Acting President of WAITRO in his welcome address said:

> All over the world, in developed as well as developing countries and in transition economies, we are seeing a challenge to industrial research and development institutes. Our economies are evolving into truly knowledge-based economies where the competitiveness of any economy hinges on its ability to make R&D at an international level available for the last company in the industrial structure of the country.— Institutes in many countries used to have a considerably easy existence based on fairly abundant availability of good and non-competitive sources of funding. However, now in all communities availability of funding is diminishing and we are thus expected to prove our efficiency and our ability to deliver in the market place (Qureshi).

Several developing countries have launched programmes

* This study draws upon heavily from the Indian part of an international study coordinated by the World Bank entitled "Policies and Institutional Priorities for Industrial Technology Development". The Indian part of the study was jointly funded by ICICI, Bombay and CIER, Taiwan and the research team was led by the author. He gratefully acknowledges the contributions of his team members, Profs. M.R. Dixit and S. Ramnarayan and the contributions of M/s. C.B. Dasgupta, Dilip Hegde, Naresh Chotai, Arijit Sikdar, Sumit Mustafi, Suresh Sharma, Drs T Rangarajan and Anil Yadav.

to restructure their technology institutions to respond to the changing needs of industries which are undergoing a process of liberalization and globalization. Sri Lanka (Jayatissa, 1996) and Malaysia (Nesadurai, 1990) among others have initiated programmes for restructuring technology institutions. In India subsequent to the initiation of the New Economic Policy in mid 1991 the CSIR launched a new strategy aimed at (a) reorienting R&D programmes towards the needs of industry, (b) forging alliances with various links in the innovation chain, and (c) effective management and marketing of its knowledge base (Joshi, 1996).

This study briefly presents the findings of a study of industrial technology development in India undertaken during the period 1993-95 (i.e. shortly after the launch of the New Economic Policy) and then discusses case studies of the experiences of several technology institutions some of which were in the process of transforming themselves during the study period. The others had already undergone restructuring programmes prior to the new policy. The technology institutions whose experiences are discussed in this study are: Institute for Plasma Research (IPR), National Chemical Laboratory (NCL), Shriram Institute for Industrial Research (SRII), Centre for Development of Advanced Computing (C-DAC), National Centre for Software Techniques (NCST) and Ahmedabad Textile Industry's Research Institute (ATIRA).

Survey Findings

Indian manufacturing firms use a variety of organizations as sources of technical know-how and services amongst which technology institutions form an important group. Technology Institutions (TIs), as defined in this study may be either profit making or not-for-profit organizations that provide technological support to industrial firms. They may be of various types: (a) educational and training institutions; (b) research organizations that include autonomous organizations, special departments and institutions directly under government ministries, private research organizations and cooperative research organizations; (c) industry-specific research and training institutions; (d) technical institutions for small industry; (e) technical consultancy organizations, information centres, (f) standardization agencies and industry associations (Chaudhuri and Dixit, May 1994).

A mail survey of 132 manufacturing firms (randomly sampled) belonging to seven industrial sectors[1] done as part of a larger study of technology development in India (Chaudhuri and Dixit, 1994) revealed that 45 per cent of the respondents had used national technological institutions as a source of technical services; 40 per cent had used local and regional technical institutions; 38 per cent had used universities; 36 per cent had used research associations and 19 per cent had used academic associations. These figures of TI usage by industrial firms in India are significant considering that the TI usage rate ranged from approximately 32 per cent for small firms to around 75 per cent for large firms belonging largely to the same industries across 8 countries that included economically advanced as well as developing ones (World Bank, 1995).

What was also interesting to note was the respondents' reported usage of foreign collaborators as a source of technical know-how and services. Only 33 per cent of the mail-surveyed firms had used foreign investors as sources of technical services. This information apparently contradicts the popular perception of the dominance of foreign firms as sources of technical know-how in Indian industry. The study covered a cross-section of seven industries spanning several firm-size categories, which perhaps explains the apparently contradictory finding.

I must hasten to add, however, that the same respondents ranked foreign investors as "number 2" amongst the 12 sources of technical services in terms of their importance as a source of technical services. What one may surmise from this is that though a relatively small percentage of firms had used foreign investors, those that used them found them to be very useful. Incidentally firms' own R&D departments and laboratories were ranked at the top as sources of technical services on both counts, which is understandable given the tacit nature of much technical knowledge and the control

1. The industrial sectors included auto parts, foundry, machine tools, pharmaceuticals, polymers, textiles and software.

that a firm can exert on the development and deployment of its technological resources.

As is evident from the above, on an aggregate basis, the usage rate of technical services supplied by TIs is not too low though there is ample room for improvement. Let us take a look at the nature of services used by the firms. Table 1 provides information on ranking of services on the extent of usage.

From the ranking of services we find that standards and testing, education and training, and information services are the most used services. Commercial advice, collaborative R&D and contract R&D are the least used services.

This pattern of TI service usage by firms as evident from the mail survey is corroborated by an interview survey of firms (purposively sampled to ensure a spread of firm size and variety) in the same industries. The interview survey revealed weak R&D collaboration between firms and TIs. Only 20.5 per cent of interviewed firms collaborated with industry associations, 14.5 per cent with research associations, around 8 per cent with general R&D institutions, 14.5 per cent with specialized R&D institutions and around 20 per cent with technical colleges (Chaudhuri and Dixit, May 1994).

Table 1
Ranking of Services on the Extent of Usage

Services	*Used by No. of Firms (%)*	*Rank*
Standards/testing	88 (73%)	1
Education/training	81 (67%)	2
Information services	80 (61%)	3
Problem solving/trouble shooting	75 (62%)	4
Technical networks	66 (55%)	5
Commercial advice	52 (43%)	6
Collaborative R&D	30 (25%)	7
Contract R&D	24 (20%)	8

The weak link between industrial firms and TIs in the area of collaborative and contract R&D is very evident; however, the foregoing discussion also points out the strengths of the TIs in the areas of standards/testing, education and training, information services, problem solving and troubleshooting and technical networks.

In the following sections we examine the cases of a few TIs which have been successful in developing relationships with industrial firms on a sustained basis or are in the process of doing so and finally draw some lessons for TIs.

Select Case Studies of TIs

A. Generic TI Under Government Department: Institute for Plasma Research (Chaudhuri, *et.al.* 1994)

The Institute for Plasma Research was established in 1986 as an autonomous research institute by the Department of Science and Technology, Government of India to pursue experimental and theoretical studies in the field of high temperature magnetically confined plasmas.

The Institute is active on a broad canvas of fundamental studies of the plasma state. Engineering groups skilled in technologies of ultra high vacuum, pulsed power, micro-wave and RF, computer-based control and data acquisition and computer simulation. IPR has set up the Plasma Processing Group (PPG) to focus on developing plasma-assisted material processing technologies for Indian industry. PPG interacts closely with entrepreneurs through all the phases of technology generation and transfer—from concept to commercialization.

IPR has developed India's first plasma nitriding reactor for hardening of industrial tools such as gears, cutting tools extrusion dyes. The technology has already been transferred to M/s Indian Plasma System Limited, Gandhinagar, the capital of Gujarat. IPR expects that the first commercial version of the reactor would be marketed shortly. Plasma nitriding is a process which uses only 1/8 of the energy used in conventional salt bath nitriding. It is also a very clean process as it does not use any hazardous chemicals. IPR is currently in the process of developing a number of other technologies: synthesis of aluminium, high value fine ceramic, fine metal and alloy powders, silicon dioxide deposition to produce anti-corrosive properties, barrier coatings and etching of plastics to increase printability.

Though IPR was set up as an autonomous institute in 1986, it had its genesis within the Physical Research Laboratory, a research laboratory under the Department of Space. Within PRL a group was set up in 1972 to do basic research in the area of plasma physics. In 1982, a plasma research programme was set up within PRL with the objective of expanding the scope of work to include research in the area of fusion. Finally, in 1986 PRL spun off IPR as a separate institute under the umbrella of the Department of Science and Technology.

PPG's attempts to interact with industry started with (1) dissemination of information about plasma processing and its applications through a newsletter published by the institute called *Plasma Processing Updated* which is written in a lucid manner. The newsletter describes various plasma technologies and their possible applications in simple language. The institute wanted to take up for development specific products and processes in which Indian industry had already shown an interest. Since the institution's resources were limited, the selection of the specific developmental activity had to be chosen on the basis of a careful analysis of the current and future needs of industry. To gauge industry's requirements a meeting was organized in August, 1990 at Ahmedabad with representatives of Indian industry to,—

(a) identify specific technologies on the basis of industry's needs,
(b) identify expectations of the industry from the institute in terms of R&D support,
(c) get a feel for industry's commitments to utilize the results of R&D efforts, and
(d) identify strategies for effective and mutually financial interaction between the institute and industry.

Meetings with industry were organized in four other cities—Hyderabad, Madras, Delhi and Baroda. One indication of increasing interest in plasma technologies is the fact that only 200 copies of the newsletter were printed for the first issue, whereas in 1994 IPR was printing about 900 copies per issue, which was expected to go up to 1000 in a short time. The meetings in the different cities were attended by representatives of 20-25 firms. At the time of this study IPR was planning to organize a national convention on plasma technologies for industrial use.

Plasma Nitriding Project

The entrepreneur, who had joined up with IPR for developing the plasma nitriding technology, was already familiar with the working of the Plasma Group at PRL as he had worked as an engineer in the same organization. On leaving PRL he had become an entrepreneur, manufacturing micro-wave RF connectors. This person got interested in plasma technologies through a metallurgist friend of his, who had attended the meeting organized by IPR in Baroda.

An MoU was signed between IPR and the entrepreneur. The terms and conditions of the agreement were: IPR would be paid prototype development cost, a lumpsum fee to be paid in five instalments, royalty of 4 per cent after five years. The agreement was for exclusive transfer of technology. IPR was able to develop the prototype within a year of commencement of work. The plasma nitriding reactor passed all the tests. The prototype was planned to be scaled up by IPR. Drawings of manufacturing the reactor would be supplied by IPR to the entrepreneur. Another important point to be kept in mind is that the entrepreneur had already done a market survey of the potential of plasma nitriding.

Framework for Plasma Process Technology Generation and Transfer to Industries

IPR has through interaction with industries over the last few years developed a method of working with industrial firms for the development and commercialization of plasma technologies. This is described below:

Task 1: Identification of the Required Technology

This activity is best done by the potential user of the technology, namely the entrepreneur. This may involve market survey, estimation of financial investments and other inputs, competitiveness with other processes, etc. IPR can, however, help the industries in information gathering of the technical aspects necessary to arrive at definitive conclusions of the viability of the venture.

Task 2: Validation Experiments

After a specific product or process is identified by the industry as potentially interesting, the basic process of know-how has to be developed in a series of activities of increasing technological complexity. However, very often, the confidence in the technical feasibility of the basic process itself has to be established in fairly simple experimentation. Since the entrepreneur may not have the infrastructure to take up this research and development work, IPR would be willing, under extremely fair terms to help the entrepreneur to go through this phase. The basic laboratory facility will be provided by the IPR and the industry will have to provide manpower to help the scientists undertake the R&D work. We believe that this kind of involvement is essential in this phase, as it helps the entrepreneur to develop confidence in the process while helping him to appreciate the technical problems which will have to be overcome before full commercialization of the process.

Task 3: Prototype Development

The next phase is the prototype development. This means development of a shopfloor compatible system with most of the automation and safety engineering built in, and meeting the production requirements, including volume. Since the investments will be reasonably high, the entrepreneur, if he so wishes can take help from technology promotion agencies like the National Research Development Corporation, an autonomous body of the Government of India providing venture capital for such tasks.

IPR would provide the technical support for the detailed engineering design, fabrication supervision, etc., to meet the required technical specifications of the prototype system. In addition it will work closely with the entrepreneur in the commissioning of the system.

Laboratories under the Council of Scientific and Industrial Research (CSIR)

B. National Chemical Laboratory (NCL) (Ramnarayan, 1994; Lalkaka, 1996; and Krishnan, 1996)

There are around 40 laboratories under the fold of CSIR. A few such as the NCL, Pune, have achieved world class in a few select fields. The laboratory, established immediately after independence, has the mandate to — conduct research in different areas of chemistry, develop new technologies to utilize the country's natural resources, help in import substitution and increasing export potential in the areas of its interest and assist industry in technological development. The current thrust areas of the institution are catalysis, biotechnology, organic chemical technology, polymers and other high performance materials, and basic research in chemistry and biochemistry.

NCL has achieved considerable success in interacting with industrial clients. One measure of its performance on this front is the ratio of external earnings to total income. In 1994-95 NCL's external earnings amounted to 36.6 per cent of the total income. This, however, was achieved over a considerably long period of time.

During the first phase of its evolution from its inception till the mid-1960s NCL emphasized basic research reflecting the orientation of the directors and scientists who had previously worked in universities.

In the next phase the focus shifted to research on short-term applied research with a view to developing processes for industrial chemicals in keeping with the government's economic policies that encouraged import substitution. NCL introduced policies for (a) identification of projects, (b) internal coordination, (c) planning and programming, (d) techno-economic evaluation of projects, and (e) transfer of technology to industry. During the third phase beginning with the mid-1970s, it started emphasizing the need to accomplish major breakthroughs, and efforts commenced to achieve a blend between basic and applied research. NCL achieved a breakthrough in the field of catalysis with the commercialization of the Encilite-1 catalyst in collaboration with IPCL in 1985. By the end of 1987-88 the laboratory's external earnings stood at 18.9 per cent of total income. The pressure to increase earnings from services rendered to industry was strengthened by the recommendations of the Abid Hussain Committee which suggested that the CSIR finance a third of its expenditure through sponsored research.

Several major changes came about after 1989 when a new leader, Dr R. A. Mashelkar, currently the Director General of CSIR, took charge as the director of the laboratory. The new direction that he attempted to give to NCL could be encapsulated in two key concepts: (a) international competitiveness in technology development and (b) excellence in basic research.

The major steps taken by NCL since 1989 that have had a considerable impact on its performance vis-a-vis relationship with industry are briefly described below:

1. In 1990, the Project Planning and Development Unit (PPD) was set up under the director to give a thrust to NCL's interaction with industry and to achieve internal consistency between different activities. PPD's functions included: (a) gathering information on clients' technological needs and conveying the same to scientists; (b) arranging meetings between firms and appropriate divisions within NCL; and (c) acting as the commercial agent for NCL.
2. The same year a new scheme was started to support fresh' ideas on a competitive basis.
3. In 1991, the NCL Research foundation was set up with financial contributions from industry to provide recognition and give cash awards to outstanding groups and individual efforts in the development and commercialization of technology.
4. In 1992, NCL negotiated a loan from the World Bank to upgrade its infrastructure including pilot plant facilities.
5. NCL established relationships with multinationals like General Electric and Du Pont. These have helped the organization to (a) upgrade its experimental techniques, (b) improve its knowledge of patents, and (c) improve project review techniques.

6. In 1993, NCL received a grant from the World Bank to commission a consulting organization, Generics of the UK, to help in developing a management system to integrate its R&D activities with business strategy and marketing. The grant also covered training of personnel. A budget and expenditure control system was designed and implemented. Authority for incurring expenditure was delegated to project leaders and a project record management system was introduced.

C. Private Research Organization : Shriram Institute for Industrial Research[2]

The Shriram Institute for Industrial Research is a private research institute founded by late Lala Shriram, one of the well-known industrialists of the country. The institute has had a very chequered history. It was founded in 1944 and was headed at that time by Dr S.S. Bhatnagar, a very well-known Indian scientist, who was later to become the head of the CSIR. The institute had a very interesting genesis. Lala Shriram felt that for the economic development of the country it was essential to build up scientific and technological skills. However, he felt that the research should be directed towards industry's needs, which he thought could be achieved only if it was set up with the help of funds contributed by trade and industry. He himself gave Rs. 9 lac and also some free land for the setting up of this institute and was able to mobilize some additional funds from a few other industrialists. The institute was named after Lala Shriram by the shop-keepers, who had together given him whatever they could to establish this institute.

The period 1974-79 was very painful for the institute. It did not have funds to pay salaries to the employees, who became unionized. The situation became so difficult that the CSIR was on the verge of taking over the institute. However, negotiations fell through. At that time the late director Dr Nigam, who was to become the director working in a private sector firm. He was invited to head the institute as the joint

2. Based on discussions between the author and late Dr Nigam, the then Director of SRII.

director. The new director was given full reins to turn around the institute.

A number of actions were taken by Dr Nigam. It was decided that anybody who was identified as surplus in the organization would have to leave. Perquisites enjoyed by senior scientists had to be given up. The organization decided not to use any peons, sweepers, security-men, gardeners, typists, etc. A major decision was taken to employ scientists against retirement of people at any level. For routine jobs it was decided to recruit graduates instead of people with masters' degree. In 1979, out of 300 people in the organization 75 per cent were supporting staff and the remaining 25 per cent were scientists having masters and doctoral degrees. By 1994, the ratio had completely changed. At the time of this study 75 per cent of SRII's personnel were scientists and only 25 per cent were supporting staff. The sales turnover of the organization in 1994 was around Rs. 6 crore, which was a major jump from a turnover of Rs. 50 lac in 1978.

The services which SRII provided in 1994 included information services, toxicity analysis, testing services, and project irradiation. The institute was in the process of setting up a computer-aided design centre in Bangalore. Another unit for manufacturing ultra-pure chemicals required in micro quantities was also being set up. These ultra-pure chemicals are very high-valued items sold in quantities of grams. The philosophy of technology development of the organization is "not to do technology package development rather to fill up gaps."

D. Sector Specific Institution : Centre for Development of Advanced Computing (C-DAC) (Chaudhuri and Dasgupta, 1995)

This institution was established in 1988 to design, develop and deliver supercomputers by the Department of Electronics (DoE). The reason for setting up this institution was that the technology for supercomputers was not available to the country from abroad, thus hampering the development in various fields (e.g. meteorological forecasting). It is a generic type of technological institution set up as a scientific society in line with the Centre for Development of Telematics (C-DoT) thereby giving it greater autonomy in its decision-making than the CSIR laboratories. But all the R&D activities of the TI were funded by DoE as its output was considered to be of national importance and because of its large requirement of funds, which was unlikely to be funded by private parties. Part of the funds came from contract R&D and manufacturing. Thus government influence existed in funding support and determining the broad research and technology agenda.

The mission of the institution was to develop super-computers indigenously for which it was involved in the development of parallel processing hardware, software and techniques. For realizing its mission the institution was provided with the latest facilities and equipment like Very Large Scale Integrated Circuit (VLSI) manufacturing process and electronic hardware fabrication equipment.

The institution's technical activities were directed towards strategic R&D with focus on the industry's future needs than on basic and academic R&D. This was visible in the greater importance attached to practical achievements in evaluating the performance of its technical staff and negligible importance attached to academic achievements (publications). In working towards its mission the institution kept itself abreast of the latest technological developments in the world which could be put to use by it. Its practical achievement orientation was evident in its technology acquisition/development strategy whereby it first tried and used existing solutions, failing which it attempted to get technology from outside and finally attempted to develop technology on its own. This was due to the fact that the institution's mission was to get technology faster to the user. This strategy paid off. The institution was able to develop a state-of- the- art parallel computer PARAM in the first 5 years of its existence. It was also able to transfer GIST products, for multilingual technology for commercialization.

Out of its total 160 staff, about 110 were technical staff whose average age was low. This was its greatest asset. The success of the institution lay in its management and culture. A problem solving approach with its researchers given leads to follow as well as the practice of assigning specific study tasks ensured that projects were well directed. The informal work culture cut down the bureaucratic wrangles and thereby ensured timeliness. Pride in their achievements and the knowledge that they were the technological "torch bearers" motivated the staff in their efforts. Also motivation was derived from the emphasis given to practical achievements in evaluation.

For developing its internal capabilities the institution placed emphasis on "homegrown" technology based on worldwide technological trends. Technology induction was mostly through study of literature. For training its own staff the institution had its own training school—Advanced Computer Training School (ACTS). It also had a lot of interaction with external institutions, both public and private, which acted as co-developers. It used private firms for manufacturing and marketing facilities which it did not have.

It also had a lot of interaction with its clients whose feedback it took in developing its next generation of products. To be more market-oriented the institution was planning to use market research to determine its products.

E. National Centre for Software Techniques (NCST) (Chaudhuri and Dasgupta, 1995)

This institution was established in 1985 as a national software laboratory by the Electronics Commission. It had its genesis as a group set up within the Tata Institute of Fundamental Research (TIFR). The TIFR environment in which it was born provided it with the culture of accountability, international orientation and autonomy, which other labs lacked. Government grants constituted 30 per cent of the income, while the rest came from the sale of services. The primary missions of the institution were to carry out research and development in software technology, undertake educational and training activities and support industry through R&D. The organization had a strong application orientation towards industry while basic research was not given much importance. This was evident in the evaluation of technical staff where more importance was given to practical achievements than academic ones. It defined its role as a leading edge assimilator of technology for its industrial clients.

To keep up its technological level it acquired the latest facilities in technology in terms of computing hardware, communication and library facilities. To update the technical capabilities of its personnel the institution organized training under UNDP fellowship for specific projects, sent people to international computer manufacturers for training and arranged courses, workshops and conferences in India by invited experts.

To keep itself abreast of the latest technological developments the institution collaborates with various universities in India and abroad for joint research, conferences and workshops and, with R&D oriented companies like Computer Maintenance Company and Tata Consultancy Services, it collaborates for joint technical activities. Also to develop internal capabilities the institution has created a centre at Bangalore, created a division in software engineering, entered into MoUs with universities, both Indian and foreign, and companies from abroad for technical and scientific cooperation.

For the future the institution does not see a change in its objectives but believes that a higher level of technical support needs to be provided to clients which necessitates that it should invest more in R&D and maintain a high level of effectiveness of R&D so that it can benefit from the technological development. The institution was found to be quite successful and was referred to by most of the software firms for their technological needs the low publicity that the institution enjoyed notwithstanding.

F. Cooperative Research Association—Ahmedabad Textile Industry's Research Association (ATIRA)[3]

ATIRA owes its origin to the vision of two industrialists of the city of Ahmedabad, who were able to mobilize financial support from the mill owners' association in the city and that of the Central Government. They were also successful in securing the support of some influential scientists. Conceived

---

3. The case study is based on published information, annual reports, and discussion with senior scientists of the organization.

as an industrial research organization to subserve the technological needs of the composite mills in the city it has undergone a major change in its *raison detre* because of the dramatic changes in the structure of the Indian textile industry. From the very beginning ATIRA's basic objective has been to apply the methods of science and technology to improve the capability and output of textile mills. The specific objectives have, however, changed over time depending on developments in its external environment.

Has it been successful in achieving its basic objective? To answer this question is not easy. What criteria should be used to measure the performance? In the absence of clearly identified and measurable criteria one may consider several important aspects of the organization in assessing its performance: growth in membership, services rendered to industry, performance in the area of research and development, financial performance, and self-renewal capability.

Growth of membership is an important measure of performance. It is a measure of the faith of industrial organizations in a technology supporting institution. ATIRA has been fortunate to have experienced steady growth in its membership. However, one needs to dig a little deeper to understand the dynamics. The original profile of ATIRA membership has undergone considerable change. Whereas in the beginning the entire membership consisted of composite textile mills, today the founding group constitutes only about 35 per cent of its members. Does it mean that the original founders have withdrawn support due to dissatisfaction with the services provided by ATIRA? It may partly be true for some of the organizations. But a large majority resigned from their membership of ATIRA because of the financial difficulties they were facing.

In the initial years services did not account for any significant income. However, over a period of time ATIRA has been very innovative in responding to the emerging needs of member mills. From the very beginning the management of ATIRA considered developing a close relationship with its member mills — a critical activity. Hence the activities that they undertook from the very beginning were designed in such a way that they required constant interaction with member mills. From almost the very beginning ATIRA started bringing out on a monthly basis abstracts of latest developments in textile science for the benefit of the member mills.

One must remember that the concept of industrial research organizations was very new when ATIRA was set up. Therefore, it was a very uphill task that ATIRA had to undertake in winning the industry's confidence. In the beginning the technical services provided by ATIRA included routine physical and chemical testing, assessment of work load, training for quality control, etc. Over a time ATIRA's scientific and technical personnel have acquired competence to provide a variety of services to the industry and its policy-making bodies. From time to time it has attempted to provide new services in accordance with the emerging needs of the industry whilst at the same time it has discontinued not the so popular ones.

The problems tackled by ATIRA as a part of its consultancy services have varied from transient shopfloor problems to surveys of mill operations, planning of renovation, modernization, expansion and planning of new mills, comprehensive performance analysis, optimization of performance of various departments, energy conservation, etc.

In recent years consultancy in the areas of pollution control, defect analysis and development of process control systems for minimising value loss, assessment and calibration of expensive laboratory instruments, and customized information packages, have been introduced. Other services include inter-firm comparisons of financial performance, costs, productivity, staffing patterns, yarn and fabric quality, etc. A variety of training programmes for different levels have been held in order to update knowledge and competence and promote a professional culture. Workers, jobbers and supervisors are given attitudinal training and demonstrations of proper work methods with respect to their levels. Organizational restructuring and top management role clarification also form a part of its capability portfolio.

ATIRA has also conducted programmes for technologists and top management on technical subjects using a variety of pedagogical methods.

ATIRA has developed a comprehensive range of testing services for fibres, raw materials, intermediate products, and finished goods. It has created a specialized laboratory for testing selected machinery accessories and spares.

Product and process development has been an important activity at ATIRA. A variety of instruments, machinery accessories and control systems have been developed which can be used directly by the industry. More than 40 such developments have been licensed to the industry.

Research and development at ATIRA covers a wide range of activities from basic research to applied research and development. Oriented basic research has been considered very important by ATIRA as it can lead to a better understanding of the properties of textile materials and also help in injecting rigour and sharpness in to applied research. Basic research covers fibre physics, textile chemistry, chemistry and chemical modification of starches and gums. Most basic research at ATIRA has been made possible through extramural funds provided by the CSIR and other national institutes, the US Department of Agriculture and other international organizations. As a result, it has tended to decline in times of resource scarcity in the funding organizations. The relative decline in the importance of basic research has been alluded to in the previous sections. One indicator of this is perhaps a greater emphasis over the last two decades or so on recruitment of technologists/engineers with Bachelor's or Master's degrees rather than Ph.Ds. from top world class educational institutions as in the initial years of ATIRA's development. Of course, this change of orientation has its good obverse side when looked at from industry's perspective. ATIRA over the years has become more attuned to industry as evidenced by the nature of work done and the consistent increase in the earnings from various kinds of services.

The issue of balance between basic and applied research and development is very critical for an organization like ATIRA. Prolonged de-emphasis of basic research, which ostensibly started as a result of financial difficulties, may have a debilitating effect in the long run. There is already a feeling amongst the larger and relatively more sophisticated textile firms that they do not gain much from organizations like ATIRA as they feel they themselves are technologically at a higher level. This may or may not be true. But the perception is a reality. Unless ATIRA is able to strengthen its activities supported by generation of new knowledge it would tend to become less relevant to the technology oriented larger firm's. The longer this situation continues the greater would be the challenge to bring about a radical change in orientation.

Over the years ATIRA has exhibited an ability to manage its operations in a condition of financial stringency barring short periods of relative abundance of financial resources when major infusions of funds were received. In the area of financial performance ATIRA's experience has varied over its history. During its early period ATIRA was dependent to a great extent on grant in-aid from the Central Government though industry support mainly through membership subscription was also substantial. A regime of financial stringency began in the early sixties when the support from the government was reduced and become uncertain. Fortunately other international agencies stepped in and provided support through research sponsorship.

ATIRA responded to the situation by using a variety of actions: not filling vacancies, cost reduction, partial reorganization and enhanced attention towards income generating activities like consultancy and testing services, sponsored research, and licensing of new product and process know-how. These steps helped ATIRA diversify its sources of income considerably and weather the storms it faced. One document — 40 Years of ATIRA — brought out in 1987 stated that over the period 1972-87 membership subscription accounted for 28 per cent, government grants 28 per cent, sponsored research projects 21 per cent and services rendered 23 per cent. These ratios have undergone further changes. By 1992-93 membership fees constituted 17.80 per cent of the total income, government grant 25.30 per cent, sponsored research 31.9 per cent and income from services and licence fees 23.80 per cent.

These figures speak for themselves. ATIRA has successfully diversified its sources of income and decreased considerably its dependence on government grants. Not only that, it has been successful in generating a significant amount of funds through specific projects tailored to customers' needs and through services rendered.

ATIRA has also exhibited a capability for self-renewal. Beginning with a strategic planning exercise in 1974, which involved only a few selected persons, it has undertaken several comprehensive reviews at 5-year intervals which resulted in the redefinition of its philosophies, objectives and strategies. Actions stemming from these comprehensive reviews have enabled ATIRA to respond to major structural transformations of its client industry.

Implications for TIs

What lessons can we draw from the previous consideration? Obviously, drawing lessons from a disparate set of cases is fraught with many difficulties. Nevertheless an attempt is made to discuss some implications for TIs that emerge from the cases.

1. The Scientific and technological institutions in India are passing through a difficult period due to the financial stringency imposed on them as a consequence of the government's adoption of the new economic policy in mid-1991. For many the challenge of raising a third of their revenues from client-sponsored projects is a tall order. But it is evident from the success stories of some TIs that developing relationships with industrial clients is not an impossible task. Success has been achieved by a variety of TIs operating in high tech environment in catering to industries based on moderately sophisticated technologies. The TIs themselves were of various types; ranging from fundamental research institutions to cooperative research associations focused on an industry. Successful TIs seem to develop strategies that are unique to their situations. There does not exist any one best method of developing partnerships with an industry. However, there seems to be some common elements across the case studies of effective TIs. These are identified and discussed in the following paragraphs.

2. Ownership does not seem to be a critical variable explaining TI success in building sustained long-term interaction with client industries; however, it might influence

the character of interaction. NCL is high on science, technology and international competitiveness; its motto is 'a deal a day' with an emphasis on short-term problem solving and provision of services that enjoy economies of scale; while ATIRA's focus is on a mix of activities including technology development and providing short-term services for process optimization, productivity and quality improvements with the balance tilting towards the latter. Evidently the three TIs had very different orientations. NCL is a governmental laboratory under the CSIR; SRII is a private research organization and ATIRA is a cooperative research association; the latter two having strong industry orientations. It may be surmised that a strong representation by industry in the committees responsible for developing TIs' research, technology development and service provision plans might be a useful organizational practice for facilitating partnership with industry.

3. A paradigm shift from offering readymade technologies to a collaborative mode of identifying client needs, developing and transferring of technologies is likely to improve the effectiveness of TIs.

4. For TIs new to the game of developing partnerships with industry it may be useful to initially focus on short-term problem-solving, process optimization, training and education and information service. This would enable the TI to get an entry into the technology service market. Gradually with better understanding between the TI and the industry and a higher credibility for the former it may be possible for it to move into more sophisticated areas like collaborative and contract R&D.

5. In situations which demand a missionary orientation, radical changes in policies, major changes in an organization's vision and orientation or response to apparently stupendous odds in technology development as exemplified by the cases of ATIRA, NCL, SRII and C-DAC, the quality of leadership is of the utmost importance. The role of leadership is critical in articulating a clear mission for the TI, communicating and imbibing the organization with values consistent with its mission, mobilizing and motivating key constituencies, building the necessary capabilities and developing and introducing necessary management systems.

6. Transition from a technology push paradigm to a collaborative technology development paradigm would require TIs to acquire skills in marketing, interacting with industry, generating and managing funds in a more sophisticated way; instituting new planning and control systems; building a new culture that emphasises achievement of results rather than adhering to rules and procedures, etc. TIs would have to invest in training and development activities in various areas to acquire the new skills.

7. In the long-term TIs would need to arrive at a balance between serving the industry's immediate needs and building capabilities for meeting the industry's future requirements for new product and process development. In the context of an increasingly performance-oriented environment that is simultaneously characterized by considerable uncertainty TIs may benefit from using formal strategic planning to evolve their objectives and strategies.

REFERENCES

Chaudhuri, Shekhar and Dasgupta, C.B. "Exploiting the Potential of the Indian Software Industry", Working Paper No. 1267, June, 1995, IIM, Ahmedabad.

Chaudhuri, Shekhar and Dixit, M.R. *"International Study on Policy and Institutional Priorities for Industrial Technology Development: India", Vol. 1: Integrated Country Report*, (IIM, Ahmedabad), 1994.

Chaudhuri, Shekhar and Dixit, M.R. "Institutional Infrastructure for Industrial Technology Development in India", Working Paper No.1186, May 1994, IIM, Ahmedabad.

Jayatissa, J.M., "Industrial Technology and Support Services to SMEs — Experiences of CISIR, Sri Lanka", in M.A. Qureshi (ed.) *Human Resource Needs for Change in R&D Institutes* (World Association of Industrial and Technological Research Organizations and NISTADS), 1996, pp. 29-37.

Joshi, S.K. "R&D Strategy of Council of Scientific and Industrial Research, India", in M.A. Qureshi (ed.), *Human Resource Needs for Change in R&D Institutes* (World Association of Industrial and Technological Research Organizations and NISTADS), 1996, pp. 5-7.

Krishnan, Rishikesha T. Response of Natinal R&D Laboratories to Environmental Changes : A Study of Five Indian Industrial R&D Laboratories, FPM Dissertation, March, 1996, IIM, Ahmedabad.

Lalkaka, Rustam, "Applying Technology for Enhancing Competitiveness" in M.A. Qureshi (ed.), *Human Resource Needs for Change in R&D Institutes* (World Association of Industrial and Technological Research Organizations and NISTADS), 1996, pp. 39-65.

Nesadurai, H.E.S. "Government Management of Research and Development in Malaysia — Responses to New Economic Imperatives", in Tarek M. Khalil and Bulent A. Bayraktar (eds), *Management of Technology-II — The Key to Global Competitiveness* (Proceedings of the Second International Conference on Management of Technology, Miami, Florida, USA), 28th February-2nd March, 1990, pp. 239-251.

Qureshi, M.A. (ed.), *Human Resource Needs for Change in R&D Institutes* (World Association of Industrial and Technological Research Organizations and NISTADS), 1996, vii.

Ramnarayan, S., "Tough Challenges Facing the Indian Polymer Industry: Unshackling the Follower Mindset", in Shekhar Chaudhuri, M.R. Dixit and S. Ramnarayan *"International Study on Policy and Institutional Priorities for Industrial Technology Development: India"*, Vol. 2: Sectoral Studies, IIM, Ahmedabad, 1994.

The World Bank, "Institutions and Policies for Industrial Technology Development", November, 1995.

8

Management of Technological Innovation in a Public Funded R&D Laboratory System

Santanu Roy

Research on different aspects of invention and innovation have continued to remain a matter of debate for a long time. Nelson and Winter (1977) suggest that there is a need for a reconciliation between studies aimed at pinpointing sources as invention, and those concerned with the diffusion of innovation. The subject of success/failure of the technological innovation process has remained a thrust area of research from the time of early pioneers like Mansfield (1968) and Burns and Stalker (1968). Freeman (1982) defined innovation as the technical, industrial and commercial steps which lead to the marketing of new manufactured products, and to the marketing of new technological processes and equipment. Five similar government R&D laboratories were studied using Critical Functions Analysis by Rhoades *et al.* (1978) to determine critical functions and a profile was developed for each laboratory. Sherlock and Edge (1977) have made a case history study of the R&D department of the ICI Agricultural Division. Wagner (1982) has studied General Motors of the USA and Toyota of Japan and have delineated the success or failure in innovation by the large organisations. Knut (1987) has probed into user-oriented product innovation.

From a review of project termination decisions and

discussions with several R&D managers, Balachandra (1985) presents a set of four critical factors which can lead to certain termination of the project if there is a large unfavourable change in any one. These are the probability of technical success, continued existence of the anticipated market, government regulations, and raw material availability. This study identifies a set of discriminating factors and variables based on a study of over 50 R&D projects in the electronics, data processing and communication (high technology) industries. The key variables are incorporated into a scheme for monitoring the progress of projects in the R&D portfolio. In another study, Balachandra (1996) makes a comparison of R&D project termination factors in four industrial nations — Japan, the United States, Germany and the United Kingdom.

Apparently, there can be ambiguity in determining whether a project is a success or a failure. There are two main reasons for this ambiguity. First, as mentioned in a study by Pinto and Slevin (1988), it is still not clear how to measure project success because the parties who are involved in projects perceive project success or failure differently. A project which is considered to be a success by the client might be considered a failure by top management if the project outcome does not meet their management specifications, even though it might satisfy the client. In this case, both the parties are evaluating project success differently and thus they value the outcome differently. The second reason is that lists of success or failure factors vary in various studies in the literature. Although several lists of factors are generated, they seem to tabulate individual factors rather than group them according to some criteria, to help analyse the interaction between them and the possible consequences. Furthermore, many of these factors do not, in practice, directly affect project success or failure. Usually a combination of many factors, at different stages of a project life cycle, result in project success or failure. There might be additional problems with the use of these tables. For example, due to the unique nature of projects, most of the factors in a list might not be applicable for a particular project, or a factor which is the main determinant of success for a project might not be listed (Belassi *et al.*, 1996).

Significantly, the body of literature cited above has tried to delineate the factors responsible for success/failure of technological innovation efforts retrospectively. Rickards (1985) suggests that the term "innovation" should retain the recognition of the importance of modest improvements in organizational practices. This distinction has various, very practical consequence. If we do not consider incremental change to be essentially innovative, we eliminate it from our consideration, and any search for 'best practice' will concentrate on rather rare cases. Furthermore, focussing on "big bang" innovations overlooks the possibility of innovations in processes, which in turn may lead to innovative outputs.

There has been little attempt to probe into the strategies adopted for managing technological innovation and the implications of such strategic choices made upon the overall thrust of the innovative efforts. The present study puts forward a strategy initiative for the management of technological innovation by drawing lessons from the technological innovation efforts pioneered by a public-funded R&D laboratory system, the national laboratories under the Council of Scientific and Industrial Research (CSIR).

The Strategy

What is proposed is a strategy initiative in the context of management of technological innovation. The R&D/production interface could be defined as a temporal, spatial and functional concept which represents the organizational and interpersonal (managerial and professional) exchange of information and know-how. It is also a part of the R&D / Innovation spectrum which can be: an arena for co-operation or conflict; a generator or repository for ambiguity about the locus of responsibility; and a source of high costs, delays, and technical difficulties. When the interface works smoothly and quickly, products and processes seem to "slide" effortlessly through the R&D/innovation process. When it does not, the process falters and often breaks down (Rubenstein and Ginn, 1985). The strategy initiative proposed in this study calls for consensus among the actors in the innovation process and a vision to guide the process.

It calls for a holistic approach to look at the issue in its totality and identify the actors in the innovation process right at the outset who would all work towards a common objective.

The present study examines this aspect in terms of a few minimum strategic initiatives. The study refers to case studies of technologies developed by three laboratories of CSIR. It, however, does not attempt to detail out the case studies as such but attempts to draw lessons from these studies relevant to the appreciation of the strategy initiatives being proposed here.

The present scheme, therefore, visualises this strategy as being composed of the following primary initiatives:

(I) Vision

A strategy for technological innovation must have a vision to begin with, an appreciation of the situation and the problem at hand and the need to resolve the same. Identifying the actors in the process (members of the consortium or the network) to achieve the common objective is crucial to this phase as also an appreciation of the likely role to be played by each one of them in the innovation process. Carrying out techno-economic feasibility studies and other pre-project studies is important. Arriving at a consensus among the actors and developing strategies for cooperation begin here.

(II) Regional Specialization

Though this need not be true in every case, yet regionalization of specific technological innovation efforts has become institutionalized in India in certain cases. One could see the concentration of chemical industries in western India or of information technology groups dominant in parts of southern India. What is being proposed is not only a recognition of this phenomenon but promotion of such endeavours.

(III) Consortium Mode

The importance of identification of different partners in the innovation process and the likely role to be played by each one of them in the process cannot be underestimated.

By a consortium or a network we refer to an alliance of interested members who all have a stake in the outcome of the effort and who are in the process of making a joint bidding for achieving a common objective. This is to be arrived at through a process of mutual consensus among the actors in the innovation process. Each actor attempts to fulfil the mutually-agreed task assigned to him, always recognizing and appreciating this particular role in order to accomplish the common goal of the consortium.

(IV) Design Engineering

The aspect of design engineering has been observed to occupy a very significant and critical position in so far as the success of technological innovation efforts in the Indian situation is concerned. Concentrating on a flowchart of a technology from the bench scale to the laboratory scale to semi-pilot plant/pilot plant scale and finally to the commercial scale, one finds that this flow is not only non-linear but also the non-linearities are often so complex as to render all the basic work done by the laboratory scientific team uselass even with the skeletal design engineering groups a few laboratories possess.

The above four premises do not constitute a complete strategy set but only a minimum set of strategy initiatives as a first step of conscious action.

The role of the State or the government in this scheme is crucial and perhaps very critical. The State could act as a promoter, a facilitator or a provider in this strategy initiative. This could take many forms like enactment of new legislations, statutes or providing venture capital or funding board for new innovation initiatives, spreading awareness through government-controlled or otherwise electronic or other media and above all be receptive to such moves or proposals. In other words, the State would facilitate the formation of a network among the concerned actors for achieving a common objective. The exact nature and scope of such a role would vary with the situation prevailing at that point of time.

In the light of the above strategic initiatives for management of technological innovation and formation of a conscious network among the actors the innovation process being the driving force behind such an effort, it would be pertinent to look a bit more closely into the Indian situation with a view to learning from these experiences as also to appreciate the significance of adopting such a strategy initiative.

The Case Studies

In each of the case studies, an attempt is made here to identify the different components of the strategy enunciated above. The case studies, when subjected to critical dissection, throw up instances or situations which serve as learning experiences in our present framework. These are, therefore, not detailed accounts of the projects' flowcharts but are helpful in so far as they bring out certain issues relevant to the present study.

A. Sponge Iron Production (National Metallurgical Laboratory (NML), Jamshedpur)

The objective of the programme was to produce sponge iron—almost completely metallised lumps by direct reduction (DR) of iron ores involving in situ reduction or iron oxide without involving fusion. NML has developed two processes for the production of sponge iron using non-coking coal as fuel and reductant—the rotary kiln process and the vertical retort furnace process.

(I) Vision

The project work was initiated in the background of the following:

(a) There was a gap between the demand and supply of steel.

(b) A large number of electric steel plants using non-coking coal were expected to come up which are less capital-intensive as compared to conventional steel plant and which do not require coking coal (it was expected that the production of steel by electric steel furnaces would approximate to three million tonnes) and sponge iron would serve as a feedstock for these plants.

(c) The technology was solid reductant based as compared to natural-gas based DR processes in other countries in view of the limited suply of natural gas in India.

(II) Regional Specialization

NML was set up as a metallurgical laboratory in the coal producing belt of Bihar and the adjoining state of West Bengal, as coal is the basic fuel in most of the metallurgical processes. NML's research themes had as one of its thrust areas the exploitation of low-grade, high-ash non-coking coal, abundantly found in this belt but unsuitable for use in the conventional blast furnaces. The particular process of sponge iron production had also been developed keeping in mind the need to exploit the supply of non-coking coal as a fuel and as a reductant. However, this regional specialization of the technology and of the laboratory itself was of little practical significance as there existed a parallel government policy in vogue for a long time, which stipulated that the price of coal would be uniform throughout the country, denying any particular advantage which might have accrued due to the regional specialization aspect of the process.

(III) Consortium Mode

The two technologies developed had followed different courses of action—the consortium mode was conspicuous by its absence for the earlier technology but was present in the latter case. The earlier process was the rotary kiln process which had not been tested adequately in the pilot plant scale before it was transferred to M/s. Andhra Cement Company, Vijayawada, where it was tried at their old rotary kiln after modifications. However, it could run for only about six months and even during that period, there were many breakdowns. On the other hand, the vertical retort furnace technology was a distinct improvement in that it offered better volume utilization and other advantages. But more than that, a pilot plant was designed and built in the laboratory itself (initially a 350 kg per day capacity where metallization went up to 90-95 per cent). For the first time, process utilization of the furnace gas was achieved in this pilot plant which boosted investor confidence in no small measure. Consequently, the commercialization of this technology was achieved in collaboration with M/s. Mechanical Engineering Consultants, (MECON), Ranchi and a pilot plant of 12-15 tonnes per day capacity was set up with a capital outlay of Rs.70 lakhs, being shared jointly by MECON and CSIR.

It is to be noted that NML's R&D programmes were oriented to suit the requirements of dynamic growth and expansion of Indian mineral land metal industries under the impact of successive five-year plans. It becomes imperative in pursuing various applied projects to study on pilot plant scale trials, potential practical themes so as to determine their suitabilities for commercial exploitation under Indian raw material conditions. This resulted in progressive establishment of various pilot plants. However, with the shift in CSIR's policies towards pilot plants and with the stipulation that pilot plants need to be industry-sponsored, the pilot plants slowly met their final end. There was no concerted effort from the laboratory's side or even from the side of CSIR to evolve a policy of involving industries in their R&D work so that it would be in everybody's interest to support the pilot plants which are a vital link between the laboratory scale operation and the commercial scale operation, especially for large scale or process-intensive technologies.

(IV) Design Engineering

The earlier technology of rotary kiln process was transferred to M/s. Andhra Cement Company, Vijayawada where the design of the plant was carried out by the laboratory scientists themselves without any external support in terms of expertize. Moreover, a forty-year old cement kiln was converted to function as the plant. Having failed in this attempt, for the vertical retort furnace technology, NML not only had the advantage of having a full-scale pilot plant working in the laboratory itself (where a critical process engineering problem of utilization of furnace gas was resolved), but also went for a collaborative effort with MECON, Ranchi in the commercial phase. This design engineering consultancy firm had the role of designing and developing the engineering specifications for its commercialization. A modular plant with four retorts (capacity

12-15 tonnes per day) was established jointly funded by CSIR and MECON for this purpose.

B. Chlorosilanes — National Chemical Laboratory (NCL), Pune

Organo-silicon polymeric compounds "silicones" are industrially very important compounds with many applications. The work was taken up as a basic research initially and it took an applied course afterwards:

(I) Vision

Silicones have multifarious applications such as in grease, paints, rubbers, resins, adhesives, emulsions, etc. Silicones possess a unique combination of metallic and non-metallic properties. This technology was developed in the West, but it was kept a closely guarded secret by the multinationals. Indian companies were importing intermediates for the manufacture of these products. This was, therefore, essentially an import-substitution technology.

(II) Regional Specialization

NCL is strategically located in the western part of the country where chemicals and other allied industries abound. The success of this technology in particular and of NCL in general are in no small measure related to the regional aspect of industrialization of this nature.

(III) Consortium Mode

NCL management took a conscious decision to collaborate not only with a user company (M/s. HICO Products Ltd.) but also with a process engineering firm (M/s. Humphrey & Glasgow Ltd.) right at the very outset of the applied R&D work (formation of the consortium). M/s. Humphrey & Glasgow Ltd. had carried out a techno-economic feasibility study and costing of the project. The design engineering problems were resolved using the expertize of this firm.who are a specialist in this area at the beginning of the pilot-plant work (the level of technology transfer). In fact, the final project manual was prepared jointly by the three parties. Further, close monitoring of the progress of the project and the high level of active cooperation and coordination among the user of technology and the generator of technology were also crucial to the success of this effort. Standards were fixed, and the process and product specifications conformed to it.

Another instance could be cited as an illustration of the significance of this consortium mode strategy initiative. When NCL developed the chlorosilane technology, import of this technology was still continuing. The matter eventually came up before the Technology Evaluation Committee (TEC) of the Directorate General of Technology Development (DGTD). It goes to the credit of NCL, HICO and National Research Development Corporation (NRDC) who put up a joint fight in all the available forums and were ultimately able to stall the import of this technology which at that point of time was extremely vital for the technological innovation to succeed.

(IV) Design Engineering

The technology being process-intensive, there were a few critical elements in the process engineering aspect which required to be resolved. Some of these were,—

(a) the recovery of methyl chloride (an intermediate) has to be as high as possible as it is recycled in the process and governs the economy of the process;

(b) the design of the reactor for the highly exothermic reaction (taking care of temperature, pressure, catalyst, heat control, etc.).

Thus, this particular process is engineering-intensive involving complexities of gas-solid reactions and possesses typical scaling-up problems from the laboratory/pilot plant scale (up to 3.5 kg. per hour) to the commercial scale at HICO Products Ltd. The process engineering specialists, M/s. Humphrey & Glasgow Ltd., along with NCL and user company were instrumental in designing the commercial scale plant, thus demonstrating the success of this innovation strategy.

C. Mini-Tiny Cement Plants using Vertical Shaft Kiln Process —10 to 100 tonnes per day ([Regional Research Laboratory (RRL)], Jorhat)

The technology developed was for the production of ordinary portland cement of IS:269 quality of the Bureau of Indian Standards (BIS).

(I) Vision

Here vertical shaft kiln technology has been adopted in mini/tiny scale holding the advantages of utilization of smaller deposits of limestone, lowest capital investment per unit capacity and low gestation period. The vision was to disseminate this technology far and wide, adaptable to the local conditions prevailing in different parts of the country. It was meant to promote and decentralize an industry very basic to our infrastructural development. There are about twenty running units spread all over the country based on this technology.

(II) Consortium Mode

The project team had carried out a detailed analysis of cost and profitability of 25 tonnes per day capacity plants. Consultants and fabricators are appointed under a tripartite agreement for offering the technology on a turn key basis, perhaps one of the very few instances as far as CSIR is concerned. So here the consortium consists of RRL and CSIR, NRDC, the consultants and the fabricators and the user agencies. The consultants and fabricators are jointly endorsed by RRL and NRDC. Five consultants and five fabricators have been approved so far.

It is the conscious attempt to forge such an alliance that is noteworthy in this case. The consultants in collaboration with the fabricators render a variety of services to the entrepreneurs which include the following:

(a) arrangement of licences;
(b) preparation of feasibility reports;
(c) arrangement for evaluation of raw materials;
(d) assistance in securing finance from financial institutions;
(e) preparation of detailed project engineering report;

(f) technical evaluation of tenders and vendor analysis;
(g) advice on selection of site;
h) inspection during civil construction;
(i) installation and commissioning of individual plants, machinery and quantity of product;
(j) arrangement for training plant operating personnel;
(k) plant commissioning including stabilization for rated quality and quantity of product;
(l) assistance in providing BIS certificate and other technical assistance as and when required.

(III) Design Engineering

This point has already been stressed on in the earlier section. The design engineering for the pilot plant in the laboratory was carried out by the scientists themselves. But in the technology transfer phase, this factor was adequately taken care of by the consultants and fabricators endorsed by RRL and NRDC.

Discussion

The above three case studies provide learning situations to appreciate better the significance of the strategy initiatives proposed in this study. For the sake of brevity, only three case studies have been considered here. The analysis of these cases lends support and reinforces the proposed scheme or strategy initiatives for a better management of technological innovation.

The near absence of any financial institutions in the consortia studied above is a very serious drawback and requires careful consideration and focussed attention. This could take care of the financing of pilot plants, provision of venture capital and other financial necessities. The recent attempts at forging an alliance between CSIR and ICICI (Industrial Credit and Investment Corporation of India) could be a pointer in this regard.

What is required is the establishment of a network among the participants in the technological innovation process.

A network comprises all the organizational or social units connected by a specific type of relationship. Conceptually, a network may develop at various levels. These levels include intergroup, interdepartmental, interorganizational, and international. A particular network may depend on one or more of the following types of relational content: (a) communication content — the exchange of information among entities; (b) exchange content — the flow of goods or services among units; or (c) normative content — shared expectations that network member units have of one another based on some social feature. Networks also may develop to enable members to actuate increased political, social, or economic power and to legitimate new social structures. Thus networks may vary greatly depending on purpose or function as well as level (Chisholm, 1996).

According to Trist (1983), interorganizational networks perform three basic functions. The first function is regulating — assuring appropriate orientation, value maintenance and organization of the network. Regulating contrasts with operations that are carried out by stake-holders. Developing a shared appreciation of a desirable future and of trends and issues that emerge over time comprises the second function. The third function involves providing infrastructure support through expertise in organizing, maintaining and managing the network.

Historically, firms organized research and development (R&D) internally and relied on outside contract research only for relatively simple functions or products (Nelson, 1990). Today, these various types of interfirm alliances take on many forms, ranging from R&D partnerships to equity joint ventures, collaborative manufacturing and complex co-marketing arrangements. The most common rationales offered for this upsurge in collaboration involve some combination of risk sharing, obtaining access to new markets and technologies, speeding products to market and pooling complementary skills (Kleinknecht and Reijnen, 1992; Hagedoorn, 1993).

A key finding from a diverse set of studies is that the R&D intensity or level of technological sophistication of industries is positively correlated with the intensity and number of alliances in those sectors (Freeman 1991; Hagedoorn, 1995). In a study of network of learning in biotechnology, Powell *et al. (1996)* argue that when the knowledge base of an industry both complex and expanding and the sources of expertise are widely dispersed, the locus of innovation will be found in networks of learning, rather than in individual firms.

The case studies presented in this study clearly illustrate and reinforce the basic strategic premise of institutionalizing a network among the various participants in the innovation process. This is distinct from a cooperative research framework which is perhaps more relevant in high technology and newly emerging areas like biotechnology, as pointed out by Powell and others. In the case studies, there was no conscious attempt to institutionalize such a network, yet success was achieved when such a network got established, even temporarily. In the sponge iron case of NML, Jamshedpur (vertical retort process), the crucial design engineering and subsequent technology transfer problems were resolved through a collaborative effort of CSIR and MECON, Ranchi. In the chlorosilanes project of NCL, Pune, a network was established among the laboratory, the user company (M/s. HICO Products Ltd.), and the design engineering consultancy firm (M/s. Humphrey & Glasgow Ltd.) to tackle the design engineering and commercialization problems of the technology. Later, the network was enlarged to include NRDC to put up a joint fight in the Technology Evaluation Committee of the DGTD to stop the import of this technology crucial for the success of the technological innovation effort at that point. The network mode was also evidently responsible for the commercial success of the mini/tiny cement plants technology, of RRL, Jorhat, the network being comprised of RRL and CSIR, NRDC, the consultants, the fabricators and the user companies. The emphasis, therefore, should be given on establishing such a network consciously. A strategy for the management of technological innovation process requires recognition of the roles played by different actors in the innovation process and consciously forging an alliance among these actors in a network. The cases in this study were drawn from a public-funded R&D laboratory system but the lessons are significant and are of universal relevance.

To conclude this study, an illustration would be in order. A technology developed by the Central Salt and Marine

Chemicals Research Institute (CSMCRI), Bhavnagar, could be considered. The technology, "Recovery of Potassium Chloride from sea bittern through the Medium of Mixed Salt" was potentially very promising. India currently imports about 20 lakh tonnes of this chemical only for the fertilizer industry, and it is also very important for the defence sector. With India possessing no potash deposit, economically recovering this chemical from sea bittern (the mother liquor left after salt production) is perhaps the only way out. However, the project was only partially successful (the only notable successful transfer in the past being to the Gujarat State Fertiliser Corporation). This was the case in spite of the fact that the laboratory is strategically located in Bhavnagar (in the salt-producing Saurashtra region). Apart from innumerable design engineering problems faced during the first technology transfer to M/s. United Salt Works, Kandla (there was no specialized design engineering support), the most serious drawback of the entire innovation planning was the near absence of the consortium mode strategy initiative. For the recovery process to be economically viable, perhaps the only feasible solution is to recover it as a by-product from existing salt firms (along with recovery of chemicals like sodium sulphate and magnesium sulphate). The laboratory went in for developing the technology for the recovery process and tried to transfer the same to the fertilizer companies or salt works which depend upon the salt firms for the raw material (bittern). The investment required is quite large, and the process becomes uneconomical as a commercial proposition in most of the cases. What was really required was the involvement of big salt firms possessing the required capacity and investment capability to go in for by-product recovery right at the initial stages of the innovation effort. That was not the case earlier, and in the Kandla plant of M/s. United Salt Works, apart from the process engineering problems (like the faulty design of the cooling tower which was later rectified by incorporating atmospheric spray cooling), the plant had to be shut down as it was not proving to be feasible to run it commercially. In fact, CSIR had appointed a consultant, M/s. Industrial Consultants Bureau, to suggest possible alternatives for reviving the Kandla plant and make it a commercially viable unit. Such an effort did not bear any fruit.

The study puts forward a proposal for adopting some strategy initiatives as a first step of conscious action. The study argues that an established network of concerned actors working towards a common objective could achieve a better management of technological innovation by drawing lessons from cases of technological innovation efforts pioneered by a public-funded R&D laboratory system.

REFERENCES

Balachandra, R., "A Tracking Model for R&D Projects", in *Project Management: Methods and Studies*, Burton V. Dean (ed.), Elsevier Science Publishers B.V., North-Holland, 1985.

Balachandra, R., "A Comparison of R&D Project Termination Factors in Four Industrial Nations", *IEEE Trans. Engg. Mgmt.*, 43 (1), 1996.

Belassi, W. and Tukel, O.I., "A new Framework for Determining Critical Success/Failure Factors in Projects", *Int. J. Project Mgmt.* 14, (3), 1996.

Burns, T. and Stalker, G.M., The Management of Innovation, Tavistock, London 1968.

Chisholm, R.F., "On the Meaning of Networks", *Group and Org. Mgmt.*, 21, (2), 1996.

Freeman, C., *The Economics of Industrial Innovation*, Frances Pinter, London, 1982.

Freeman, C., "Networks of Innovators: A Synthesis", *Res. Pol.*, 20, 1991, pp. 499-514.

Hagedoorn, J., "Understanding the Rationale of Strategic Technology Partnering: Inter-organizational Modes of Co-operation and Sectoral Differences", *Strat. Mgmt. J.*, 14, 1993, pp. 371-85.

Hagedoorn, J., "Strategic Technology Partnering During the 1980s: Trends, Networks and Corporate Patterns in Non-core Technologies", *Res. Pol.*, 24, 1995, pp. 207-32.

Kleinknecht, A. and Reijnen, J.O., "Why Do Firms Co-operate R&D? An Empirical Study", *Res. Pol.*, 21, 1992, pp. 347-60.

Knut, Holt, "User-Oriented Product Innovation in Theory and Practice", *Technovation*, April, 1987.

Mansfield, E., *Industrial Research and Technological Innovation*. Norton, New York, 1968.

Nelson, R.R., "US Technological Leadership: Where Did it Come

From and Where Did it Go?", *Res. Pol.*, 19, 1990, pp. 119-32.
Nelson, R.R. and Winter, S., Towards a Useful Theory of Innovation, *Res. Pol.*, 6, 1977, pp. 36-76.
Powell, W.W., Koput, K.W. and Smith-Doerr, L., "Interorganizational Collaboration and the Locus of Innovation: Networks of Learning in Biotechnology", *Admin. Sci. Quart.*, 41, 1996, pp. 116-145.
Rickards, T., *Stimulating Innovations*, Pinter, U.K., 1985.
Rhoades, Richard G., Edward B. Roberts and Alan R. Fusfeld, "A Correlation of R&D Laboratory Performance with Critical Functions Analysis", *R&D Mgmt.*, Vol. 9, No.1, 1978.
Rubenstein, A.H. and Ginn, M.E., "Project Management at Significant Interfaces in the R&D/Innovation Process", in *Project Management: Methods and Studies*, Burton V. Dean (ed.), Elsevier Science Publishers, B.V., North-Holland, 1985.
Sherlock, E. and Edge, H.A., "Case History Studies in ICI Agricultural Division, Research and Development Department, 1968-1975". *R&D* Mgmt., Vol. 8, No. 1, 1977.
Trist, E.L., "Referent Organizations and the Development of Inter-organizational Domains". *Human Relations*, 36, (3), (1983).
Wangner, Eugene G., "Success or Failure in Innovation by the Large Organisation — General Motors vs. Toyota", *IEEE Eng. Mgmt. Revi.*, 10, (3), 1982.

9

Institutionalization of R&D and Strategic Cooperation in Indian Paper Industry

Tabassum Jamal

Institutional basis of cooperation amongst the undertakers of R&D development, production and trade is gaining importance, which is reflected in the emergence of several cooperative arrangements world over. Transition from a "technology push" paradigm to a "collaborative technological development" paradigm is becoming a welcome concept in Indian industries as well.

The present study provides a detailed picture of such 'collaborative technoligcal development in Indian paper industry', which has emerged through strategic cooperation among the research institutions, production units and their related associations.

Section 1 of the study highlights the salient features of the paper industry, comprising major constraints faced by the industry. A focus has been made on the newsprint sector which is one of the important and crucial sectors of industry where the dependency is higher on imports. How to increase the domestic production is a big question posed by the industry today, which can only be solved by searching non-conventional raw materials particularly the agro-based plus the scope of waste paper utilization.

The awareness and realization of other important parameters like energy consumption, chemical recovery, manufacturing of paper machinery and automation augmented the establishment of public-funded R&D institutions. Section II highlights the role of public-funded research institutes and other in-house R&D divisions of the large paper mills to cater to the industry requirements. Technical and business associations like Indian Pulp and Paper Technical Association (IPPTA), Indian Agro Paper Mills Association (IAPMA), and Indian Paper Makers Association (IPMA) are detailed in Section III. These associations have been a major source of strategic cooperation brought to industry by providing the information to the mills, on the latest government policies and negotiating the industry's terms and demands with the government on industry's behalf.

Salient Features of the Paper Industry

Paper manufacture in India started in the beginning of 20th century, and today we have 380 paper and paper board mills (Table 1) with an aggregate capacity of nearly 3.8 million tonnes, with the capacity utilization of 74 per cent during 1994-95. There are 7 newsprint mills producing 350,000 tonnes of newsprint annually. Our per capita consumption is still very low at 3.5 kg/annum against a world average of 4.7 kg. (Rao, 1994)

There are three distinct segments of the Industry, namely forest based, agro based and 'others' principally based on waste paper and secondary fibres. Presently, forest based units constitute 43 per cent of capacity, agro-based units 28 per cent and others 29 per cent. Forest based units are mainly located in the eastern and southern zones of country while 'agro' and 'others' are concentrated in the northern and western Indian States.

India's demand of paper and newsprints is projected to grow from the present level of 4.5 million tonnes to 6.9 million tonnes by the year A.D. 2010. Present pattern of consumption, comprising 55 per cent as cultural varieties and 45 per cent as industrial and packaging varieties (Kraft papers and boards), is foreseen to undergo a shift towards the latter in tune with the structural changes in our economy. (Keshwani 93).

Table 1
Growth of Indian Paper Industry

Year	No. of Installed Mills	Production Capacity (Lac tonnes)	Capacity	Utilization %
1950-51	17	1.37	1.16	85
1960-61	25	4.00	3.45	86
1970-71	57	7.68	7.58	99
1980-81	135	16.50	11.12	67
1990-91	325	33.04	23.00	69
1991-92	326	33.64	25.00	74
1992-93	340	35.51	25.40	71
1993-94	372	37.90	27.30	72
1994-95	380	38.40	28.60	74

Source: Development Council for Pulp, Paper and Allied Industries, (1995), Government of India.

Structure of Industry

Large paper mills in India are invariably based on bamboo, wood and other forest raw materials, although there are some wood based mills also which have a capacity of less than 33,000 tpa. Also in recent years, large agro based mills are being set up and several of the small agro based mills are gradually expanding their capacities to 33,000 tpa and above. As raw materials are generally classified as : (i) forest based, (ii) agro based, etc., paper mills may be categorized into (a) forest based, (b) Agro based and (c) others — principally based on waste paper and secondary fibres. Units mostly using waste paper and imported pulp are classified as 'other' units and these are usually small capacity mills. However, most agro based and forest based units also use varying quantities of waste paper and/or market pulp in their furnish.

As an alternative to depleting forests, causing scarcity of soft wood to the paper industry, and in view of the abundance of agricultural residues, the government encouraged the setting up of small mills based on agro-residue like wheat and rice straws, bagasse, grasses, cotton linters, rags and industrial jute waste. At present the agro-based paper mills contribute about 36 per cent of the total production in the country. These units contribute strong waste water because of no chemical recovery and the magnitude of pollution is nearly equal to that contributed by large mills. It is estimated that a paper mill without recovery plant based on agro-residues with an installed capacity of 30 tonnes per day has a pollution load of four times that of a conventional paper mill with chemical recovery and installed capacity of 1000 tonnes per day. (NPC, 1989)

Newsprint Sector

There are 27 manufacturing units in the country, making for a total capacity of six lakh tonnes. Of these, five public sector units, viz. Hindustan Newsprint, Hindustan Paper (Naagaon and Cachar units), Mysore Paper Mills, Tamil Nadu Newsprint and National Newsprint and Paper Mills, NEPA account for 63 per cent of total capacity (Table 2). Wood based units constitute 58 per cent of capacity while agro and waste paper based units account for 23 per cent and 19 per cent of the newsprint capacity respectively. All public sector units, with the exception of Tamil Nadu Newsprint and Papers and Mysore paper Mills, (bagasse-based) are wood-based. (in-Paper International, 1996)

Table 2
Production Capacity of Large Paper Manufacturers

	Installed Capacity
1. NEPA Ltd. Nepa Nagar, Madhya Pradesh	88,000
2. Mysore Paper Mills Ltd., Bhadravati, Karnataka	75,000
3. Hindustan Newsprint Ltd., Kottyayam, Kerala	80,000
4. Tamilnadu Newsprints & Papers Ltd., Trichy, Tamil Nadu	50,000
5. Sri Rayalseema Paper Mills Ltd., Kurnool, A.P.	20,000
6. HPC Ltd., Assam	20,000
7. HPC Ltd., Cachar, Assam	20,000
Total	3,53,000

Domestic newsprint production is around 3.5 lakh tonnes per annum and capacity utilization is also satisfactorily high at over 90 per cent. As against this, the annual demand for newsprint is growing at a compound rate of about six to seven per cent. Accordingly, demand for newsprint is estimated to go up from six to seven lakh tonnes in 1996-97, and as high as 20 lakh tonnes by the turn of the century. This estimate is on the basis of annual compound growth rate of five per cent in the number of literates and even higher expected growth in the stock of highly educated manpower. This will eventually have a favourable impact on the circulation of dailies and other newspapers.

The various development aspects of the newspaper industry will thus greatly hinge on the adequate availability of newsprint. This is because newsprint accounts for as much as 70 per cent of the variable cost of the newspapers. Two important issues will determine the success or otherwise of the anticipated growth prospects. First, augmenting of newsprint availability through stepped-up domestic production, which means stepping up the existing indigenous capacity. Second, easing of import restrictions on newsprint in the context of the latest industrial liberalization trend. In either case, the important factor that will determine the growth prospects of the newsprint industry will be the price factor.

The shortfall between requirement and availability of newsprint is placed at about three lakh tonnes. If the current level of capacity remains unchanged, the demand supply gap is expected to go up to 5.5 to 6.0 lakh tonnes by 1994-95. The leading newsprint producing units, the National Newsprint and Paper (NEPA, capacity 88,000 tonnes), Hindustan Newsprint (80,000 tonnes), Mysore Paper Mills (75,000 tonnes) and Tamil Nadu Newsprint (50,000 tonnes) are all in the public sector — central or State. In addition, Rayalseema

Paper Mills has a capacity of 20,000 tonnes. With the completion of the NEPA mills modernization, the indigenous capacity has gone up. NEPA mill has been entrusted with the responsibility of implementation of the UP bagasse-based newsprint project with a capacity of 88,900 tonnes. Besides the UP project, a paper-cum-newsprint project of Punjab Agro Industries Corporation (capacity, 33,000 tonnes newsprint and 33,000 tonnes paper based on bagasse) is also in the pipeline.

This means that if these projects are implemented speedily, it is possible for the indigenous capacity to go up to around nine lakh tonnes in the next five years or so. As is well-known, newsprint is not only capital-intensive but the plant also has a gestation period of about four to five years. In respect of resources, the public sector newsprint units are not in a position to generate sufficient resources for ploughback to augment capacity, whereas the private sector investment is difficult to come by because of low returns. Further, in view of the difficult position of forest-based raw materials, the future capacity would have to be based on bagasse and other agro based raw material. Besides this there is a need to focus on efficiency improvement in the management of energy and environment as well.

Newsprint imports have risen tremendously though indigenous newsprint capacity, and output too has gone up over the years. Production went up from 0.45 lakh tonnes to 3.36 lakh tonnes over a period of ten years. As requirements also have grown at a rapid rate in the wake of a sharp jump in the number of periodicals and magazines, sizeable imports have had to be resorted to. Against around 2.0 lakh tonnes of imports during major part of the first half of the eighties, the same is in the range of 3.0 lakh tonnes now (Table 3) (Kale, 95).

Table 3
Import Dependence of Indian Paper industry

Year	*Total import value of newsprint, waste paper and market pulp in million rupees*	*Estimated imported quantities in '000 tonnes of*		
		Newsprint	*Waste paper*	*Market Rupees*
1990-91	4540	-	-	252
1991-92	3980	291	-	135
1992-93	4090	296	-	137
1993-94	4940	288	-	119
1994-95	5100	310	-	-

Newsprint import is expected to double by 1995-96 on account of the anticipated demand growth rate of about 6 per cent per annum in the coming years. While the value of newsprint imports is presently Rs. 250 to 300 crores, the same will thus be more than double in the next four years. Till 1991-92, newsprint import was channelised through the State Trading Corporation (STC). The STC's selling price of imported newsprint was based on the pooled prices of import originating from different countries, including commission, handling charges, etc. However, according to the newspaper industry, the measure of channelisation introduced in 1974-75 never worked satisfactorily at any time. The industry has, therefore, been vociferously demanding 'total decontrol and dechannelisation of newsprint. (Economic Times, 1992).

Major Constraints Faced by the Industry

As stated above, the industry is plagued with the problems relating to shortage and high cost of cellulosic raw materials. Apart from this problem the main reason for the low capacity utilization was the near absence of investment in the modernization or upgradation of technology and machines. Some machines are as old as 50 years and many of them are more than 30 years old. Machine building capacity has been very limited since the very beginning of this industry. Till the early sixties, the entire machinery was imported. Even small equipment like chest agitator and stock and chemical pumps were imported. Only in the late 60s manufacturing of some need based spares started in places of imported equipments that existed in various mills.

Inception of Tata Johnson, Utkal Machinery and finally Jessops made a brilliant start in the country in the manufacture of pulp and paper machinery in the sixties and mid-seventies. Machano Paper Machines Ltd. entered into machinery manufacturing in the early eighties and is presently doing well. Several engineering units have also entered the field around the same time but they are catering mostly to the small mills. (DSIR, 1990)

The very permission to import second-hand plant and machinery further eroded the growth of indigenous machinery builders. This has caused a setback to the industry because, with second-hand equipment at low price, the mills automatically import an outdated technology. Such imports have widened the technology gap between India and other developed countries. The best solution that can be adopted under such circumstances is to survey the machine, analyse its deficit and bring in the costs of updating the machine by incorporating latest technology to achieve the product in a cost-effective manner.

Secondly, Indian machinery builders must enter into technical partnership with world leaders abroad like Valmet, Sulzer, Steinbeis, Beloit and a few others. The Indian partnership should be technically well equipped to accept the technology transfer, absorb it and also be able to translate it on Indian conditions.

Reviewing the status of paper industry in general, one could cite the following constraints:

- poor availability of forest based fibrous raw material;
- high capital cost;
- high input cost;
- interrrupted and erratic power supply;
- technological obsolescence;
- installation of inferior and inadequate equipment in old paper mills;
- absence of investment in modernization or upgradation of technology and machines;

- lack of established and viable chemical recovery system in small mills leading to pollution and loss of valuable materials.

Keeping in view the above constraints, the industry requires an intensive research effort to enable itself in productivity and profitability. Research institutions and in-house R&D units of the mills are required to take a lead in promoting the concept of concentrating research on high priority topics of relevance to clearly identifiable targets. Being one of the oldest industries of India, the industry is privileged to have technical and business associations which have been of utmost importance to the mills by providing them information on latest government policies plus negotiating the industry terms and demands with the government on industry's behalf.

The major areas which need focus are improving productivity by increasing recycling through cleaner production methods, and improving efficiency in the management of energy and environment. All this has to be achieved through a competent, skilled and capable manpower base. Section II highlights some details as to how the Institute of Paper Technology (IPT) is catering to this need of industry.

Institutionalization of R&D

Institutionalization of R&D is a recent phenomenon in Indian Paper Industry. Realization of strategic problems (as stated above) cautioned the industry to put intensive R&D efforts in high priority areas. As a result, at present there are more than 30 R&D units engaged in pulp and paper research. Forest Research Institute (FRI), Dehradun, Institute of Paper Technology (IPT), Saharanpur, and Central Pulp and Paper Research Institute (CPPRI), Saharanpur, are the main public funded research institutes engaged in basic and applied research in pulp and paper. Indian Institute of Technology, (Bombay), Karaikudi, Tamil Nadu (Hyderabad), Central Electro Chemical Research Institute (CEERI), Birla Research Institute for Applied Sciences (Magda), Pulp and Paper Research Institute (Jakaypur) and a few other in-house R&D divisions of large paper mills such as Tamil Nadu

Newsprint Limited (Madras), Pudumjee Pulp and Paper Mills Limited (Pune), Seshasayee Paper and Boards Limited (Madras), ITC, Bhadrachalam, etc., are actively engaged in bringing research and development in Indian paper industry. The following section details how IPT and CPPRI and some in-house R&D units have assisted the paper industry by conducting research in priority areas.

(A) Institute of Paper Technology (IPT)

Established in 1963, the Institute of Paper Technology (located at Saharanpur) is the first and only institute, conducting multi-disciplinary courses in the field of pulp and paper technology at higher levels.

The Institute of Paper Technology offers academic programmes which are essentially specific. In order to make these programmes meaningful and to ensure a well-trained product, it is imperative that the institute has a very strong interface with the user pulp, paper and allied industry. The institute has made concerted efforts to strengthen this interface from many angles. These can be broadly classified as under:

(a) Involvement of industry professional in teaching and examination work.

(b) Involvement of faculty in students in integrated mill training programmes.

(c) Involvement of a senior industry leader in helping the institute in its plans for growth through bodies like Advisory Committee and Board of Studies.

(d) Involvement of faculty in industry-based organizations and associations like Development Council for Pulp, Paper and Allied Industries, Indian Pulp and Paper Technical Association (IPPTA).

(e) Involvement of faculty in industry-based national evaluation activities like National Productivity Awards, Energy Conservation Awards, formation of national standards through Bureau of Indian Standards, Central Pollution Control Board, etc.

There is a continuous effort at all levels to make the academic programmes industry relevant and towards this aim, suggestions from the industry are invited through every interactive forum and are later converted into plans of action. This is evident from the high level of support received by the students in the form of scholarships, awards and campus interviews as well as the invitations to faculty for visiting industry for specific assignment.

The Institute of Paper Technology has been active in conducting a variety of industry specific short-term programmes for national and international requirements.

The institute has embarked upon major human resource development consultancy projects for Indian pulp and paper industry. The programmes involve extensive interaction with senior managers of mills on preparation of interactive training packages. The activities include mill visits, finalising course contents, preparation of training material based on mill-specific information, delivering lectures on areas of relevance, preparation of case studies and group discussion in projects and seminars. These intensive programmes mostly at mill sites have been extremely popular and sought after. The major Indian mills involved with these programmes in recent years include Ballarpur Industries Ltd., Hindustan Newsprint Ltd., Andhra Pradesh Paper Mills Ltd., Orient Paper Mills Ltd., Hindustan Paper Corporation, Star Paper Mills Ltd., and Mysore Paper Mills Ltd. Similar tailor-made programmes have also been made for Indian paper traders and officers of government departments. (Rao, 95).

The conduct of these programmes indicates the ability and expertise of IPT faculty to organize the most demanding and effective technology programmes to the specification of the sponsoring agencies. Besides, many faculty members have been associated in delivering lectures at mill site to a number of mills. The institute also organizes many short duration continuing education programmes on a self-financing basis on topics of relevance. The faculty of such programme is drawn from industry, R&D institutions, international institutions besides the faculty of the institute. These programmes have proved very stimulating and successful.

(B) Central Pulp and Paper Research Institute (CPPRI)

Established in 1980, CPPRI is the only premier institution engaged in applied research to meet both immediate and long-term needs and to promote cooperation in research programmes with other research units. Keeping in mind the basic problems of industry, the institute has taken up research in the following areas:

1. Raw Material Evaluation : Pulping and Bleaching

Research into fibre raw materials involves pulping, bleaching behaviour of various pulps, storage and processing of agricultural residues like bagasse and straws and fibre morphology, etc.

Paper and Board Research is directed towards fibres and other raw materials and the products themselves. The work includes :

- evaluation of pulps;
- evaluation of various types of additives;
- more efficient use and treatment of recycled fibres;
- Testing of physical and optical properties of various grades of papers and paper boards;
- Fibre analysis;
- Evaluation of coating pigments;
- Printability and print quality evaluation;

2. Environmental Protection and Pollution Abatement

The focus of research is on reducing the load on the environment due to pulp and paper mill effluents. Areas of current investigations are,—

- Treatment of waste water using anaerobic treatment;
- Reduction and utilization of solid wastes;
- Development of methods for analysis of liquid and gaseous emissions.

3. Chemical Recovery and Lignin Utilization

Thrust areas of research in this field are,—

- Evaluation of physico-chemical properties of spent liquors;
- Development of suitable recovery system for small pulp and paper mills;
- Desillication of spent liquors process has been developed at CPPRI and commercially running in HNL Kerala.

4. Instrumental and Chemical Analysis

Chemical analysis of fibrous and non-fibrous raw materials is carried out using sophisticated analytical instruments.

5. Pilot Plant Facilities

These comprise of, —

- Pulp mill complete from chipper to bleaching section;
- Vertical plate falling film evaporato;
- DARS pilot plant;
- Double wire belt washer;
- Briquetting plant.

and are some of the pilot plant facilities available in the institute.

Ongoing R&D Projects

1. Refining of wood/non-wood fibres.
2. Oxygen bleaching of non-wood/wood pulps.
3. Bleaching of high-yield agricultural pulps.
4. Upgradation of quality of pulp from waste paper.
5. Studies on manufacture of highly filled paper.
6. Studies/assistance to pulp and paper mills for reducing fresh water consumption.
7. Chemical recovery.
8. Deşilication of black liquor transfer of technology.
9. Energy conservation for pulp and paper industry.
10. Development and utilization of lignin by-products.
11. Setting - up of biotechnology lab.
12. Treatment of pulp and paper mill effluents.
13. Elimination of phenolic and chlorophenolic toxic materials.
14. Gasification of waste liquor/biomass.
15. Anaerobic-aerobic treatment of effluents.
16. Jute project — under the UNDP assisted National Jute Programme being implemented by Ministry of Textiles (Government of India).
17. Solid waste reduction and utilization.

The following are some of the research projects which have improved pulping of domestic raw material, pulping of mixed agricultural residues and chemical recovery system particularly for small paper mills.

1. CPPRI's Desilication Technology

Silica is the major source of problems in agro-based paper mills and is the main stumbling block in the black liquor processing. High silica makes it almost difficult to concentrate the black liquors and to cousticise the green liquors rich in silica. CPPRI made consorted efforts for about 10 years to develop the world's first successful commercial desilication process based upon the carbonation technique to recover more than 80 per cent silica from the black liquor. A commercial plant is successfully operating at Hindustan Newsprint Ltd., Kerala, capable of processing 40 mt of black liquor per hour. CPPRI has entered into a licensed agreement with M/s ENMAS Process Technologies (Madras) to market this technology and the response from the Indian paper industry is very encouraging.

CPPRI's pioneering work on black liquor research has enormously helped the Indian recovery boiler manufacturers in designing the recovery boiler configuration. The development of desilication technology for removal of silica and development of thermal treatment for reduction of viscosity has made it possible to design tailor made recovery systems for specific black liquors. A number of companies through institutes analytical service have been able to design chemical recovery systems and three agro-residues based mills are already in advanced stages of installing chemical recovery systems.

2. Direct Alkali Recovery System (DARS)

An immediate need of small agro-residues based pulp mills is techno-economically feasible chemical recovery process. CPPRI has concentrated it efforts in this direction and carried out detailed investigations on the applicability of direct alkali recovery system (ferrite process) for agro-residues pulping waste liquors. Based on the promising results obtained on bench-scale investigations, a pilot plant has been installed and commissioned with the collaboration of a Danish company, in order to last the feasibility of the process of non-woody raw materials. This is the only pilot plant outside Denmark, capable of producing 80-90 kg. of sodium hydroxide per hour. Preliminary results obtained with pulping waste liquors of bagasse, straws have been very statisfactory with an average sodium recovery of 70-80 per cent, causticity over 90 per cent and white liquor strength up to 170 g/l as NaOH.

3. Evaluation & Upgradation of Anaerobic Technology

CPPRI is actively engaged for the last five years to execute the ancerobic treatability of agricultural residues spent liquors which is a major source of pollution in small mills. Laboratory scale anaerobic studies have been carried out in Upflow Anaerobic Sludge Bed (UASB) reactor of 30 litre capacity procured from M/s Paques B, the Netherlands. The unit comprises of pre-acidification tank/equalization of tank and the main reactor column equipped with suitable dosing pumps for feed, nutrients, ph control and temperature controllers. Laboratory studies conducted at CPPRI have revealed that agricultural residues spent liquors are analrobically biodegradable to an appreciable extent of 50-60 per cent.

Besides the above mentioned technologies developed by CPPRI, there are some other significant technologies developed by in-house R&D division of large paper mills such as:

(C) Bagasse Pulping Technology, Tamil Nadu Newsprints & Paper Ltd. (TNPL), MADRAS

TNPL developed this technology with the help of BELOIT (USA) and Seshasayee Paper and Boards Ltd., Madras. TNPL is now diffusing this technology, through consultancy services to the paper mills and turnkey project to the coming up plants. This work is being done by *Project and Consultancy Division* of the Firm which is operational for the last two years. Two examples can be cited where this division is providing all technical specifications and consultancy package to install the paper plants.

1. Sterlite Industries India Limited, Bombay

Under this project the division has prepared all technical specifications and scope for supply of continuous digesters and their auxiliaries.

2. Sri Vindya Paper Mills Ltd., Somania Nagar

The division has prepared all blueprints and specifications for installation of this plant in Ushavall, Maharashtra, which will have a capacity of 45000 TPA for producing writing and printing paper.

Collaborative Projects

The firm has also taken up the following collaborative R&D projects with various research institutes:

(a) Pulping of bagasse special sugarcane varieties, R&D projects with various research institutes.

(b) Suitability of jute sticks, jute stalks obtained from M/s Madras Gunny Exchange in the place of eucalyptus hybrid.

(c) Cause and control of corrosion in hardwood digester in collaboration with Central Electrochemical Research Institute (CECRI), Karakudi, Tamil Nadu.

(d) The suitability of bagasse washing effluent for anaerobic digestion was studied in collaboration with M/s Pudumjee Paper Mills, Pune, which showed that the effluent concentration is low for anaerobic reactor treatment.

Another example can be cited of Mysore Paper Mills taking consultancy from TNPL on machinery maintenance. As TNPL and Mysore Paper Mill are engaged basically in producing newsprint, consultancy on some common problems has helped both the mills. Such a partnership relation has helped the mills to solve the technical problems economically and comparatively in lesser time.

(D) Anaerobic Treatment System — Pudumjee Pulp and Paper Mills Ltd. (PPPM), Pune

Pudumjee Pulp and Paper Mills Ltd., (PPPM), which is the first name today in Indian paper Industry in producing cleaner technologies, has taken a lead in technology based R&D in generic technologies. The firm has improved upon anaerobic treatment system and it is in continuous operation for over six years and has already paid back capital investment. The system is generating biogas as energy source which is equivalent to about 15 per cent of the fuel oil requirements of the mill. This technology developed with Sulzer Brother is known by the name (an-OPUR-P) is available on demand. This technology could also be adopted by the existing industrial units like distilleries and food and pharmaceutical units, for waste water treatment. The mill also obtained process engineering from Sulzer Brothers for a mixed bed reactor design to give a new pulping technique, namely, solvent pulping. The process not only allows lignin recovery thereby solving the colour problem but also gives additional by-products such as furfural, and the process wastewater then needs very light treatment. This process is now proven on a pilot plant stage and detailed engineering is beginning for the installation of a 50 TPD plant at the mill site. This new pulping process, (namely PUNEC) based on solvent use, is environmentally sound and most suitable for agricultural residues like bagasse, wheat and rice straw. (Jamal, 1995).

(E) Automation and Quality Control — Central Electro-chemical Research Institute (CECRI), Karaikudi.

Electronic systems usage has been an integral part of paper manufacturing process all over. The application of electronics has helped in maintaining the quality of the paper as well as improving the production efficiency. Both in terms of reducing loss and production cost, CECRI, an establishment of CSIR, has developed an integrated electronic system which would take care of the variety and range of raw material used, capacities and speed of operation.

On a survey made it has been found that practically no Indian small mill uses any indigenous electronic system either off-line or on-line. Only a few of the large units adopt a readily available electronic system from abroad. Details in this context where it has been used and is all imported are as follows :

Head Box Consistency Monitor	- about 6 mills
Stock Consistency Monitor	- about 10 per cent of the mills
C — Frame Instrumentation	- Nil
O — Frame Instrumentation (Monitoring - 3) (Controlling - 2)	- about 5 mills

Off-line Moisture Gauge	- nil
Off-line GSM Gauge	- nil
Colour, brightness, strength	- about 30 per cent of the mills

CECRI has developed the following instruments and instrumentation systems to meet the automation demand of the industry:

1. Low consistency monitor (Locomon)
2. Viscosity transmitter (Viscotron)
3. GEM gauge
4. Moisture gauge
5. C-frame instrumentation
6. O–frame instrumentation
7. Programmeable multiloop controller
8. Brightness and colour Tester (TRICOL)
9. Paper strength monitor

Most of these instruments have been developed on the suggestion of TAPPI (Technical Association of Pulp and Paper Industry) standards and has application in small and medium scale industries.

These instruments have been put on extensive field trial in running paper mill, i.e. Seshasayee Paper and Board Ltd. Presently, technologies of this instrument/system have been taken up by a few manufacturers. This integrated electronic instrumentation will enhance (estimated by Department of Electronics) the power by 30 per cent to 40 per cent without raising a major investment. This integrated system costs around 40 lakhs and its pay back period is 12 to 18 months.

Besides the above-mentioned institutions and production units which are playing a big role in R&D industry partnership, there are many more small level research centres and consultancy agencies, catering to the need of industry. 'Chemical Recovery' and 'Effluent Treatment' are the two areas where these centres are putting tremendous efforts to bring out the viable recovery system and cleaner technologies for adoption by small and medium sized paper mills. (As it is small and medium sized paper mills contribute about 36 per cent of the total paper production in trade.) These mills, being small in nature, are not in a position to fund research programmes in these areas. The government and the financial institutions need to give them a helping hand in this respect. (IPPTA, 84)

Strategic Cooperation

The strategic cooperation which has brought consensus between the industry and government became possible with the help of technical and business associations namely. Indian Agro-Paper Mills Association (IAPMA) and Indian Paper Mills Association (IPMA), Indian Paper Makers Association (IPMA), Indian Pulp and Paper Technical Association (IPPTA) and Indian Small Paper Mills Association (ISPMA). These associations have played a pivotal role in promoting the overall development of paper industry. Following is a brief note on contributions made by each.

(A) Indian Pulp and Paper Technical Association (IPPTA)

Technical association is a professional organization instrumental in diffusing the spirit of fraternity and awareness amongst the technologists, engineers, managers and administrators engaged in the field of manufacture of pulp, paper and allied products of the country. It has been able to create understanding among the technicians that the development of the industry and their own success depends upon mutual exchange of knowledge and closer interaction in engineering and other technical matters.

IPPTA's contribution to the industry:

- Promoted the spirit of fraternity among its members to improve professional efficiency.
- Created awareness about the fast changing technology in the world for better productivity and quality.
- Provided a common platform to its members to exchange their achievements, ideas, experience and problems faced in their day-to-day professional career.

Publication and Documentation

IPPTA is the prime agency of paper industry to disseminate the latest technical information to its members through its publication and documentation facilities. *IPPTA Journal*, published quarterly, provides the latest technical information.

It also provides abstracts from journals and periodicals from a number of international technical associations of paper industry like TAPPI of USA, PIRA of UK and APPITA of Australia.

(B) Indian Agro-Paper Mills Association (IAPMA)

Indian Agro-Paper Mills Association is the apex body of paper mills based on agricultural residues like straws, bagasse, jute, jute waste, grasses, etc. The Association, which came into existence in 1987 as an exclusive spokesman of the paper mills based on agricultural residues, has helped in the promotion and the development of paper mills based on annually renewable plants. The Association has also helped to focus the attention of the Indian Government and other authorities to the importance to be attached for manufacture of eco-friendly paper and also to conserve the scarce forest resources.

IAPMA also conducts studies and training programmes on various subjects for the benefit of the agro-based paper mills.

IAPMA functions as a catalyst for the promotion and development of the agro-based paper mills. Some of the other objectives are:

(a) to undertake professional, technical and management consultation services, to undertake studies, surveys and research projects, to organize facilities for its members like testing labs and R&D centres;
(b) to collect, collate and disseminate information on agro based paper mills;
(c) to sponsor trade delegations abroad for export purpose and technology upgradation and to have interation with those connected with the industry in India and abroad, to conduct research on agro based paper and pulp;
(d) to impart education and knowledge to develop skilled manpower — the shortage of which is felt very much at present;
(e) to establish libraries and keep the industry abreast with latest technologies in the world in the field of agro based paper mill;
(f) to study and find out remedial measures in operational levels, etc.

The association, within a short span of time, has been able to bring out awareness about the need for providing policy supports for growth of paper mills based on agricultural residues in the country. The association has taken upon itself the herculean task of organizing Paperex 93 and 95 International exhibition and conference on pulp and paper industry with co-sponsorship of Ministry of Industry and other ministries, government departments along with the industry association. Paperex has been adjudged as one of the best commodity fairs organized in this part of the world.

IAPMA has also brought out a comprehensive directory of paper mills and suppliers to the mills and have plans to bring out such publications regularly.

Agro Newsletter, the official organ of IAPMA, is brought out fortnightly and contains, information of interest to the agro-based paper mills in the country. Through the medium of this newsletter IAPMA disseminates information of the paper industry in India and abroad and notifications, etc., to keep abreast the mills with the latest policies connected with the industry. The association also brings out publications on various aspects of agro-based paper mills.

IAPMA has recently brought out an international publication *Inpaper International*, highlighting the achievements of the Indian paper industry as also the economic changes that are taking place in India and their effect on the Indian pulp and paper industry.

IAPMA has conducted a number of studies in collaboration with NPC and other organizations on techno-economic feasibility of agro-based paper mills .availability and other utilization of straws and bagasse for paper mills and perspective plans for agro-based paper mills till 2000. The association has also sponsoring delegations abroad to study state-of-the-art technologies in the field of pulp and paper industry and also to see how best such technologies could be adopted in Indian conditions.

(C) Indian Paper Mills Association (IPMA)

Since its establishment, IPMA laid stress on R&D activities in high yield pulping, improving quality of products, development of new products, pulping of new varieties of non-conventional raw materials and economizing the use of utilities.

The association's representatives played a significant role in guiding R&D activities of Forest Research Institute (FRI) Dehradun and CPPRI, Saharanpur to give proper direction to its research programmes. Another significant contribution of the association was in the preparation of a glossary of terms used in paper industry and trade which is to be finalised by the Bureau of Indian Standards.

Human Resource Development is another area where the Association has been playing a significant role to be on the board of governors of the Institute of Paper Technology (IPT) in shaping its training programme and curriculum to suit the need of paper industry. Waste paper constitutes a very important item for paper making raw material in the country. The use of waste paper minimizes pollution and helps preserve wood resources. The association has been pleading with the government to exempt waste paper and pulp from levy of customs duty to make the waste paper based units economically viable. Besides this the association got the scheme of excise duty exemption for bagasse and straw extended to agro-based units.

The association has co-sponsored and actively participated in Asian Paper'92 international exhibition and conference on the paper industry in South East Asia held in Singapore in 1992. This was the first exposure given to the Indian paper industry in an international forum. The association has also sponsored PAPEREX' 93 and PAPEREX' 95.

(D) Indian Paper Makers' Association (IPMA)

From the very beginning the association has articulated various problems facing the industry in a forceful way for consideration of the authorities. The association along with IPMA has been dealing with matters such as price, quality, standards, market practices and rationalized distribution. The other issues constantly pressed on by the association in its various representations to the government are technological upgradation and revival/modernization of the existing mills for improving product quality and deriving cost benefits in manufacture and also creation of fresh capacities.

The association has always been receptive to centralized R&D activities in the paper industry and had ventilated the idea in National Policy on paper Industry submitted by the Association to the Ministry of Industry in 1980.

The association has functioned as an effective link between the government and agencies like National Productivity Council and Federation of Indian Chambers of Commerce and Industry and the member mills by collecting and furnishing data required by them on matters as varied as production, energy utilization, environmental protection, etc., in the member mills.

The association has been extending its sponsorship to various International Trade Fairs and Conferences such as EXPO PAPER ASIA 1989 and EXPO PAPER ASIA 1991, both held in Tokyo and again the PAPEREX 1993 and PAPEREX 1995 held in New Delhi to assist interaction between the industry and technology intensive companies and help buyer-seller meets.

In a nutshell, the strategic cooperation which emerges out of the activities played by technical and business associations can be categorized broadly at two levels. One is, all associations function as an effective link between the government and industry. Second is the distinct role of each association with specific objective for which the association was established. For example, IAPMA functions as a catalytic agent for the promotion and development of the agro-based paper mills, while IPPTA provides a platform to create an awareness about the fast changing technology in the world for better productivity and quality. While the Indian Paper Mills Association has a significant role in guiding R&D activities of various pulp and paper research institutes, Indian Paper Makers Association has been dealing with matters such as price, quality standard and market practices. Other associations are also putting their inputs in more intensive R&D specific areas. Still there are gaps which could only be filled with the strategic programmes/ missions to be initiated by industry itself.

Conclusions

Liberalization of Indian economy now leaves consumers with better choices as regards the quality and price of papers available to meet their needs. To respond to this competitive situation, the industry has to gear up by installing new capacity using more and more agro-residues plus by providing machinery sector with technological solutions at cost effective terms. This is an area which has never been strategised for long-term benefits. Even the big business houses like the Birlas, Thapars, Singhanias, etc., always had short-term benefit targets and the concept of "Full Set Technology" development has never been a priority. Why is it so ? The question remains unanswered. Though the contributions of public funded R&D institutions and large mills in-house research divisions have helped industry by introducing new technologies, still there is a wide gap between the two. How to fill this gap is a big question posed by the industry today. Technical and Business Associations, though their function as an effective link between the government and the industry, still needed to coordinate development policies and guidelines in a more meaningful way. Paper industry being one of the core industries, bears the impact of time-to-time dwindling policies (i.e. restrictive import policies and liberalized policies) of the government which have put the industry in an unstatic position. For a stable growth of the industry to meet the future demands, a carefully defined and mission oriented government policy is of the utmost importance today.

REFERENCES

Rao. N.J., 'Paper Production in India—An Overview' paper presented at the seminar on Strength of Environment Friendly Paper Production in India, New Delhi, 1994.

Keshwani, S.L., 'Status and Prospects of Paper Industry in India — An Overview', Paper presented at International Conference on Pulp and Paper Industry in India — Paperex-93, New Delhi.

NPC Research section Productivity, 29 (4), 1989.

In Paper International, Issue I, ½ (1), 1996.

Kale, R.B., "Raw Material Scenario for National Pulp and Pa-

per Industry up to 2010-11", Paper presented at International Seminar on Pulp and Paper Industry, Paperex-95, New Delhi.
The Economic Times - Mid week Review - *NEWSPRINT*, 9 July, 1992.
Department of Scientific and Industrial Research, Ministry of Science & Technology,
'Technology Evaluation in Pulp and Paper Machinery Industry' — A Report prepared under Technology, Absorption and Adaptation Scheme — Government of India, 1990.
Rao. N.J., 'Role of Institute of Paper Technology in Meeting Human Resource Needs of Indian Paper Industry.', Souvenir, Paperex '95, New Delhi.
Jamal, Tabassum, 'Indigenous Technological Capability in Indian Paper Industry' Rep No. 155/95, NISTADS, New Delhi.
IPPTA, 21, (2), June, 1984.

10

Role of Financial Institution for Domestic Technology Development in Industry

Sunil Mani

The research in the area of the economics of technological change has traditionally focused on attempting to provide explanations of the interactions among production techniques, prices and quantities, and on inter-industrial variations in capital intensity (Santarelli, 1995). Another area which has received much attention is the unravelling of the effect of market structure on innovation (Kamien and Shwartz, 1982). Less attention is paid by economists to analyzing the ways in which financial resources are channelled towards innovative investments at the level of both the firm and the economy.

There are essentially four impediments or barriers to domestic technology development, namely, information, risk and uncertainty, availability of risk capital and, above all, government's technology policies that may provide the correct environment for this process to flourish in a systematic manner. There are, however, no studies or surveys of this in, especially, the Indian context. Available studies[1], done in the West, have identified the lack of adequate risk finance as one of the most important barriers to this process. This could be explained in the following manner. Under normal

---

1 The most cited study in this area is by Piatier (1984) who surveyed the barriers to innovation in the European Community countries.

conditions if an activity is considered to be risky, one could attach probabilities to its potential outcomes. On the contrary, if an activity is considered to be uncertain, one cannot even attach probabilities. The technologies which are developed domestically, usually perceived by their potential users as uncertain, fall in the latter category and hence find it difficult to secure funding from the organized but the conventional capital market. This is especially so in the Indian context where typically new industrial projects have a larger debt component than equity, and the organized capital market lending by the development banks has a history of discrimination against projects based on indigenous technologies.[2]

Analytically speaking the financial requirements of technology generation are of two types (Stoneman, 1987): (1) long-term for R&D, the purchase of technology or the setting-up of mass production facilities; and (2) for working capital which is required to cover the period leading up to the time when the product is commercialized. The traditional banking system has always shied away from extending credits to R&D activities and very often the government had to step in by providing research grants and by awarding contracts. Admittedly, with government finances under severe threat, governments have accorded only a low priority to funding of R&D.

In order to better appreciate the financing requirements for innovation, economists have found it useful to identify five different stages in the growth of a new technology-based firm : R&D, start-up, risky growth, regular growth and maturity (Prakke, 1993).Table 1 identifies the activities involved in each of the stages and the likely source of such finance.

A major inference that can be drawn from the table, which is largely based on the Western experience, is the emergence of venture capital firms as a source of finance in most

2 It must, however, be stated that there is not any systematic documentation of this commonly accepted statement. The successful Swaraj Tractor case where the development bank, IDBI, is supposed to have played a crucial positive role is more of an exception than the general case. For a detailed documentation of this case see Bhatt, V.V. (1980).

of the stages. In fact, the conventional modes of financing enters the scene only from the fourth stage onwards. But in actuality that they are an important source is something that is worth exploring. This aspect is first examined against the experiences of the USA and the UK and in the next section we shall examine the Indian experience with respect to venture capital firms as a solution to the financial barrier to domestic technology development.

A point that is worth clarifying at this stage is that access to a particular source of finance to a firm is also very much a function of its size, i.e., whether it is a small, medium or large enterprise. For instance, very often, small and medium enterprises have less avenues for obtaining their requisite firms for domestic technology development than large companies for whom the conventional banking system or even internal resources can be a source. In fact, as we shall see, venture capital firms are a source of finance mainly for small and medium enterprises.

The study is structured into three broad sections. In the first section, I examine the role of venture capital as a solution to the finance barrier to domestic technology development and its actual role in two developed countries, namely the USA and the UK. In the second section its role in a large developing country, namely in India, is analyzed. The last section sums up some lessons that can be drawn from these experiences for other developing countries.

Venture Capital Industry and Technology Financing

Though informal venture capital financing has deep historical roots, it is only since the 1940s that modern venture capital, i.e., venture capital invested by specialist bodies, was developed in the USA. Most of the modern venture capital firms are privately owned though some of them receive government support.

Table 1
Stages in the Development of a New Technology-Based Firm

Stage of Development	*Activity*	*Type of Financing*
1. R&D	Feasibility Studies -technical -commercial Technical Development Market Research	Seed financing - venture capital firms (rarely) - own firms, etc.
2. Start-up	Further development setting up production and sales; reaching break-even	venture capital
3. Risky growth	Further growth; developing and second generation products	Fledgling finance (joint involvement of several venture capital organizations)
4. Regular growth	Achieving economies of scale in production and sales	Bridging finance (development) investment banks, venture capital, takeover by larger firms.
5. Maturity	Broadening technological base and management capabilities	Stock market or other exit route

Typically the venture capital investment is targeted towards new enterprises, often involving new and sometimes high technology. There are three primary characteristics of venture capital firms which make them eminently suitable as a source of risk finance:

(i) it is an equity or quasi-equity investment which means that the investor is assuming considerable amount of risk since his investments are not secured. Consequently the fund has to take the risk of failure just like other shareholders. This feature makes it different from other forms of finance such as bank loans, leasing and factoring extended by conventional financial institutions where the returns are in the form of interest and capital repayments. In this form of financing, in the eventuality of a change in the business prospects of the venture, the lender can bankrupt the borrower by calling off their loans. Thus by participating in the equity capital, venture capital firms institutionalize the process of risk-taking — a trait that is significant for successful domestic technology development. The major reward to the venture capitalist lies in the growth of the firms in which they invest ideally resulting in the sale of the firm or its floatation on the stock market[3]. The actual experience shows that this may take five to seven years from making the investment in the first place (British Venture Capital Association [BVCA]1994);

(ii) it is a long-term investment, and it is not only an investment of money but also of time and money. Investments are not made with a view to short-term profit; and

(iii) finally it is an active form of investment in which the investor also has a participation in the management of the company. This involvement will vary from firm to firm, but the majority will expect to participate through a seat on the board and thus guiding the firms on strategic and policy matters.

In short, venture capitalists generally (Coopers and Lybrand, 1994):

1. finance new and rapidly growing companies;
2. purchase equity securities;
3. assist in the development of an innovative business idea often;
4. high technology based into product or service;
5. add value to the company by active participation;
6. take higher risks with the expectation of higher rewards; and

---

3 It has been observed from the USA venture capital industry that the average holding period for the most successful investment is 7 to 10 years. The net result of this pattern is that many venture capital firms experience a loss in value during their early years, before starting to experience a steep growth in the rate of return on total portfolio. This *J-curve* phenomenon is common to venture capital firms. For details see *Venture Economics* (1988), pp. 6-8.

7. have a long-term orientation.

Needless to add, these features make it eminently suitable as a source of risk capital for domestically developed technologies. We now examine the worldwide experience of venture capital firms as a source of finance for technology development and specifically in the context of the USA and the UK (which two countries have the largest number of venture capital firms (Table 2).

Table 2
Profile of Venture Capital Firms Worldwide, 1993

Country	*First Year*	*Number of Venture Capital Firms*	*Invested Capital (US$ million)*
USA	1946	500	3071
UK	1945	119	2247
Japan	1972	80	1000
France	1975	15	770
S. Korea	1974	4	11
India	1988	10	15

Sources :

1. Ibanez. Fernan (1989), p. 22.
2. British Venture Capital Association [BVCA] (1994), p. 3.
3. National Venture Capital Association [NVCA] (1993), p. 28.
4. UNIDO (1991), p. 170.

A surprising finding is the fact that despite being a new entrant, the Indian venture capital industry is larger than S.Korea's. I now examine the evidence on venture capital firms as a source of finance for technology-oriented ventures. First the evidence from advanced countries such as the USA, UK, and western Europe in general is analyzed and this provides a background to our subsequent analysis of the Indian case. The best model of financing is a situation where much of the financing is equity-based investments in early stage technology-oriented ventures. The actual situation in a country can be ascertained by analyzing the distribution of venture capital investments according to: (a) stage-wise financing; and (b) industry-wise distribution.

The US Case

The American venture capital industry is supposed to have played an indispensable role in nurturing the growth of the country's high technology industries. The now famous American computer companies like Digital Equipment Corporation, Apple, Compaq, and Sun Microsystems are important examples of venture capital created companies. Intel in the late 1960s and Cypress in the early 1980s were venture funded start-ups in the semiconductor industry. Both Microsoft and Lotus are software industry examples. In addition professional venture capital has made it possible to create entirely new industries like biotechnology and the courier/shipping industry. Tables 3 and 4 map out the stagewise and industry-wise distribution of venture capital investments in the country.

This shows that on an average the share of early stage financing accounts for over one-third of the investments. Much of the venture capital investments seems to enter at the stage of expansion of the newly created venture. On the contrary, much of the investments were in technology-based ventures (Table 4). In fact Mowery and Rosenberg (1993) note that, the foundation and survival of vigorous new firms also depends on a sophisticated private financial system that can support new firms during their infancy. The US venture capital market played an especially important role in the establishment of many microelectronics firms during the 1950s and 1960s and has contributed to the growth of the biotechnology and computer industries. Throughout the 1970s, $100-200 million of funds annually flowed into these industries from the venture capital community, and one informed observer has suggested that by the early 1980s, flows of venture capital for high-technology firms may have been as much as $ 2-4 billion annually (p.49).

The entire investment is equity-based. Thus the current US model is one of equity-oriented financing of technology-based ventures at primarily their expansion stage, though early stage financing is still very important.

Table 3
Venture Capital Investment in the US by Financing Stage (Percentage of Investment by Value)

Financing Stage	*1987*	*1988*	*1989*	*1990*	*1991*	*1992*	*1993*	*1994*
Seed	2	3	4	3	4	3	7	4
Start-up	11	9	8	7	6	8	7	15
Other early stage	15	15	16	17	22	13	10	18
Expansion	45	40	47	52	54	55	54	45
LBO/Acquisition	20	29	21	18	3	7	6	6
Other	7	4	4	3	11	14	16	12
Total	100	100	100	100	100	100	100	100

Source:
1. Venture Economics (1993), p. 36.
2. NVCA (1994), p. 29.

Table 4
Industry-wise Distribution of Venture Capital Investment in the USA (Percentage Shares)

Industry	*1987*	*1988*	*1989*	*1990*	*1991*	*1992*	*1993*	*1994*
Technology related	71	67	64	78	84	71	70	70
Consumer related	16	11	12	10	10	8	10	10
Other product	13	22	14	12	6	11	20	20

Source :
1. Venture Economic (1993), p.28.
2. NVCA (1994), p.37.

The UK Case

The UK position with respect to these two indicators is presented in Tables 5 and 6 respectively.

Table 5
Venture Capital Investments in the UK by Financing Stage (Percentage Shares)

Financing Stage	*1983*	*1984*	*1985*	*1986*	*1987*	*1992*	*1993*	*1994*
Early Stage	27	22.4	21.9	15.3	7.2	17.9	6	5
Expansion	48.3	40.7	36.1	27.2	22.2	29	32	28
Management buy-out/buy- in	24.2	27.9	38.1	43.8	62.5	65	62	67
Other	7.2	5.2	6.5	6.2	3.1	Nil	Nil	Nil

Source :
1. British Venture Capital Association [BVCA] (1994), p. 4.
2. Mani (1994).

Table 6
Industry-wise Distribution of Venture Capital Investments in the UK (Percentage Shares)

Industry	*1984*	*1985*	*1986*	*1987*	*1992*	*1994*
Technology related	52.3	50.4	42.6	41.5	32	38
Non-technology related	47.3	47.6	57.4	58.5	68	62

Source : Mani (1994) and BVCA (1994), p. 5.

It is seen from the above tables that in the UK over two-thirds of the investment is going towards management buy-outs and buy-ins in non-technology related ventures. In fact, within a matter of two years the investment profile has undergone a radical change as indicated by the fall in the technology related investments. In short in the UK case the venture capital firms are a decreasing source of finance for innovations.

It is against this background that we analyze the role of the emerging venture capital industry as a potential solution to the financial barrier.

The Indian Venture Capital Industry as a Financial Support for Domestic Technology Development

The venture capital firms took their roots in India in the latter half of the 1980s. Within a short span of its birth however, the venture capital industry in India has started moving away from being a source of finance for commercializing technologies. As to be argued below, this state of affairs can be attributed to a great deal to the government guidelines governing its operations. Very recently the government has even repealed the guidelines[4]. The new regulations are being framed by the Securities and Exchange Board of India (SEBI) and these are expected to distance the industry away from its envisaged role as a source of risk capital for new or relatively untried or very closely held technologies which incorporate some significant improvement over the existing ones in the country. In fact, as a necessary consequence of these changes, there will be very little differences in their operations compared to the conventional development banks. Towards making these issues clearer we discuss first the origin and structure of the venture capital industry and then analyze the investment behaviour of these firms in terms of stage of financing and industry-wise distribution. The source of data for this exercise is largely based on a survey of all venture capital firms in operation in the country as of 1993 and contained in Mani (1994).

Genesis of the Indian Venture Capital Industry[5]

As seen earlier economic growth with technological self-reliance is one of the stated objectives of India's development planning though in the current context of economic liberalization one might argue that this is more of a rhetorical statement. Most firms in the manufacturing sector have relied on foreign technology imports for their technology requirements and consequently, despite the existence of a number of policies favouring domestic technology generation, its actual generation has not been much. There are various ways of measuring the rate of technology generation in manufacturing industries. One way of doing so is to analyze the expenditure on in-house R&D by firms in relation to the amount spent on purchasing technology from abroad. By comparing the two sets of expenditures we de-

4. See Economics Times, July 26, 1995.
5. This section draws heavily from Mani (1994).

fine the rate of domestic technology generating effort as follows:

Domestic technology effort of the industry

$$= \frac{\text{(R\&D expenditure)} \times 100}{\text{Direct cost of technology import}}$$

Since consistent time series data on cost of technology imports by public sector enterprises are not available especially for the period after 1984-85[6] the rate is computed for the private sector and this is presented in Figure 1.

Though a crude measure, the rate of domestic technology generating efforts in private sector industry started declining since 1982.

As part of its efforts to reverse this trend, the government enacted the R&D Cess Act 1986 prescribing a levy of 5 per cent tax on all payments for import of technology. The fund thus accumulated is to be used for setting up a venture capital fund scheme with one of the leading development banks in the country, namely the Industrial Development Bank of India (IDBI), with the ultimate object of providing risk capital to units based on domestically developed and unproved technologies. In fact, it has been shown that only about 10 per cent of what has been collected under this head has actually been transferred to the venture capital fund of IDBI (Table 7). Even the estimates of cess collected by two different agencies of the government are at variance with each other.

With the implementation of the new industrial policy (1991) much of the foreign technical collaboration agreements in the industrial sector does not require clearance by the RBI or by the Secretariat of Industrial Approvals (SIA). Consequent to this, there may not even be reliable estimates of the actual foreign exchange outgo on account of technical collaboration agreements. We also adduce evidence to show that the venture capital fund outfit of IDBI does not correspond to the definition of a venture

6. There is however, an attempt to develop a consistent time series data set on the direct cost of technology import by the public sector enterprises. See Mani (1992).

capital fund and so even the limited portion of the R&D cess that is transferred to it does not lead to the creation of technology-based ventures. In response to this, very recently the government has created a Technology Development Board (within the DST) which will hereafter be responsible for administering the R&D cess.

The actual genesis of the venture capital industry can be traced to a series of efforts by the World Bank in the 1980s (World Bank 1989) as part of its Industrial Technology Development Project in India. As part of this project a loan of $ 45 million was made available to the government to support four venture capital entities for financing technologically innovative and growth-oriented small enterprises. The government on the other hand was to re-lend this amount to four state owned venture capital firms. All these efforts finally resulted in the notification of the Venture Capital Guidelines of 1988. These guidelines govern the details of establishment, management, the nature of assistance, size of investment and the details of exiting from the ventures.[7] All the VCFs in the public sector have been established by either the banks (both development and commercial) or by other financial institutions. This is very likely to seriously hamper their role as financiers of risky projects as experience elsewhere and specifically in the USA has shown that credit analysts and lending officers steeped in a conservative financial tradition cannot be expected to become venture capitalists overnight. As mentioned earlier the venture capital association has lobbied with the government and has almost managed (pending announcement of the revised guidelines by the SEBI) to get several of the provision of the guidelines of 1988 repealed.

Structure of the Indian Venture Capital Industry

Currently (1995) there are about 11 VCFs in operation. Of them, 7 are in the public sector and 3 in the private sector. One more fund in the public sector is in the offing. Needless to add the industry is still evolving (Table 8).

According to certain estimates, the total pool of approxi-

---

7. For a detailed commentry on the guidelines, see Mani (1994), pp. 18.

mately Rs 2.5 billion is considered to be adequate. These estimates are arrived at on the basis of the number of technologies and licences transferred and contracted by the various sources of local technology. An aspect brought out by Table 8 is that, on an average, disbursements have been only 48 per cent of sanctions. The precise reasons as to why such a gap exists are not known. As seen earlier, the main return or reward to the venture capitalist is in the form of capital gains at the time of divestment. But in the budget speech for 1995-96 the finance minister had announced an exemption from tax on income by approved VCFs in unlisted companies in certain sectors. Further, an amendment has been made the SEBI Act providing it with powers to regulate VCFs. These and the other changes contemplated in the revised guidelines are to make VCFs an attractive proposition for investment and this is also expected to increase the number of entrants to the VCF industry and at the same time move it away from the original concept of venture capital funding as an effective solution to the financial barrier to innovations in the Indian context.[8] There is also considerable concentration of assistance in just three government-owned VCFs. (Fig. 2).

Table 7
Estimates of R&D Cess Collected and Transferred to the Venture Capital Wing of the IDBI (Rs in Million)

Year	*Estimates of R&D Cess Collected*		*Estimates of Foreign Exchange Outgo*		*Transferred to the VCF of IDBI*
	A	B	A	B	
1988-89	163.2	137.2	3264	2746	88.4
1989-90	272.0	223.2	5440	4464	50
1990-91	301.2	300.1	6024	6002	10
1991-92	327.7	220.4	6554	4408	100
1992-93	448.9	369.6	8978	7392	30
1993-94	607.4	490.2	12148	9804	NA
1994-95	636.6	469.2	12732	9284	NA

Note : A — Estimate by the Controller of General Accounts.
B — Estimate by RBI.

8. For a detailed commentary on the guidelines, see Mani (1994). pp. 18.

Table 8
The Structure of the Indian Venture Capital Industry (1995)

Name of Fund	*First Year*	*Size of Fund*	*No. of assisted companies*	*Total Sanctions*	*Total disbursements*	*Nature of ownership/ Spoonsor*
TDICI	1989	1000	87	414	246	Public
IDBI	1987	478	64	43	32	do IDBI
RCTC	1988	450	10	94	-	do IFCI
Gujarat	1990	240	8	70	29	do GIIC
AVCL	1990	135	7	20	5	do
Can-Fin	1989	100	27	89	50	do
SIDBI	1994	100	4	16	NA	do
Credit	1986	100	32	50.4	NA	Private
20th	NA	108	NA	NA	NA	do
Indus	1992	NA	NA	NA	NA	do
Grindlays Bank	1992	NA	20	120	NA	Private-Foreign

Source : Mani (1994), p. 20 and Ramesch and Gupta (1995), p. 23.

The dimensions of investments analyzed are: (a) the stage-wise investments; (b) instrument-wise investments; and (c) the industry-wise distribution of investment. I start with the overall trends in VC assistance (Fig. 3). Though it has grown at an average rate of over 100 per cent per annum, the average investment per firm has remained restricted to about a half a million rupees. This is partly due to the regulations imposed by the Guidelines of 1988.

The ideal form of financing is one which involves relatively more equity form of financing in the early stage of firms operating in technology-based industries. However, this model is rarely to be found in actuality. The one more close to this ideal case is the USA one. The Indian case we argue below, does show at least in its initial stage some affinity to emulate the ideal case.

Distribution of Investment by Financing Stage

The table shows that nearly two-thirds of the investment is in start-ups. But it should be mentioned that of late there has been a shift towards financing expansions by existing undertakings. Table 9 considers only the average during a period. However, not much is known about the beneficiary companies because it is estimated that it will take at least five years before it begins to show some results.

Table 9
Venture Capital Investment in India by Financing Stage

Financing Stage	*Average during 1987-93*
Start-Up	66
Other Initial	1
Expansion	33
Others	NIL

Source : Mani (1994), p. 21.

Distribution of Investments—Instrument-wise

Most of the technology-based ventures require equity support for the reasons mentioned earlier and it is the best way of institutionalizing the process of risk associated with new and untried technologies. The guidelines specify that the financial support to the enterprises should be mainly of the equity type though it does not specify the exact amount. The only specification is that the investment in any one firm should not exceed 10 per cent of the VCF, and it should be in enterprises where the total investment in plant and machinery does not exceed Rs 100 million. The VCFs are a mix of investment instruments: equity and quasi equity instruments such as convertible conditional loans and income notes. But the share of standard loans is expected to be low. However, an analysis of their pattern of assistance since inception through 1992 shows that the predominant mode of financing is through conditional loans (Table 10).

Table 10
Venture Capital Investments in India—
Instrument-wise, 1987-1992 (Rs in Million)

Instrument	*1987*	*1988*	*1989*	*1990*	*1991*	*1992*
Equity	1.05	2.13	39.22	60.03	110.32	196.13
Conditional Loans	-	52.64	319.92	171.56	157.65	171.59
Conventional Loans	-	-	319.50	15.75	31.05	42.04
Income Notes	-	-	2.6	10.49	23.90	4.50

Source : Mishra (1995), p.173.

It is essential that the share of equity support in their total assistance should be high and, if not, increasing over time. Based on this idea we construct a simple index to be termed as financial barrier index. This is defined as the reciprocal of the ratio of equity to total financing of a project, multiplied by 100. The more close the index is to 100, the lower the financial barrier and so on. I have constructed this index for the entire industry based on the data contained in Table 9 and this presented in Fig. 4.

There has been a significant decrease in the Financial Barrier Index FBI over the years indicating an increase in equity support showing thereby that the venture capital assistance has lowered the financial barrier to innovations. However, recent discussions show that the preferred mode of assistance is going to be through a variety of debt instruments. The proposed changes in the venture Capital Guidelines of 1988 are expected to remove any stipulation that the support should be mainly of the equity type. If that happens, it will be a setback to the role of venture capitalists as financiers of risky projects.

As noted in Section (i), apart from providing the pure financial support the factor that makes venture capital operations rather unique is the managerial support components— the close involvement with the management of the investee firm (Sagari and Guidotti, 1992). Through this process of involvement the venture capitalist is supposed to add value to his portfolio, which is accomplished through strategic planning, management recruiting, supplier-customer relations, support in securing additional finance, etc. Among the venture capital funds surveyed it is only TDICI which seems to provide such value adding support functions to its assisted companies. This has manifested itself in the form of managerial and technical support. A unique mechanism adopted by the fund is to arrange for an annual gathering of chief executives of the assisted companies to understand their problems and to explore ways to improve mutual interactions. We do not of course have sufficient data from the assisted companies about the real value of these exercises to them. However, there is some evidence of general satisfaction with the nature of involvement with its assisted companies (Sesham, Sekhar, 1994).

Another very useful support mechanism which the TDICI initiated was Technology Information Service. This was essentially to access information from over 800 foreign databases. The company claimed in its second annual report (1989-90) that the information services will have an important role to play in improving managerial effectiveness and competitiveness in the years to come'. However, in the very next year (i.e. 1990-91) this was dispensed with on grounds of unpopularity and also considered the continuation of the service a diversion from its main line of activity. It is a fact that most entrepreneurs and especially those operating in high-tech industries do not have easy access to major developments in their respective areas. This form of embodied technology transfer can very often act as a fillip to domestic initiatives. But with the arrival of many on-line information services like for instance the internet this may be less of a problem.

Industry-wise Distribution of Venture Capital Investment

An important characteristic of the industrial economy of most countries is that technology has become an important medium of competition between firms. Technological changes in a number of industries have reduced barriers to entry by lowering the cost of entry and have made a number of industries scale independent, enabling even small scale units to enter areas which have been the exclusive preserve of large scale units. New technologies also imply high-risk with high-risk cannot raise from conventional funding sources. The venture capital investments has therefore emerged as a solution to this problem by specializing in extending primarily equity support to growth-oriented technology-based firms.

The guidelines governing the operations of the venture capital firms are very clear on this respect. Second, an examination of the primary objective of the various VCFs shows that their investment preference is clearly for technology-based units. Table 11.

Table 11
Investment Preferences of Indian Venture Capital Firms

Name of the Venture Capital Firms	*Investment Preferences/Objectives*
TDICI	To provide finance particularly venture capital assistance to promote commercial, market oriented technology ventures and thereby accelerating, the technology development process in India
RCTC	Commercialization of substantially innovative domestic technologies by small and medium companies which lack resources to implement them
IDBI	Encouraging commercial application of indigenously developed technologies, and encouraging the adaptation of imported technology to wider domestic application
AVCL	To promote technology-innovative projects in the State of Andhra Pradesh
Gujarat Venture	To finance technology-innovative projects in chemicals, drugs, electronics, and other engineering industries in the State of Gujarat
CVCF	Commercial exploitation of the new technologies.

Source : Mani (1994), p. 26.

In fact this is brought out by the distribution of their actual investments. Table 12.

Table 12
Industry-wise Distribution of VC Investments (Percentage of Investment by Value)

Industry	*Cumulative Investments, 1988-1992*
Technology related	78
Non-technology related	22
Total	100

Source : Mani (1994), p. 27.

It is seen that much of the investments are in technology-related ventures, though it is not clear how much of this technology originated from government research institutes. In terms of type of industries, computer software and hardware, chemicals, electronics, biotechnology, speciality materials, etc., are the important recipients (Fig. 5). Interviews with the officials of the larger VCFs reveal that increasingly they tend to prefer ventures based on proven technologies which is another way of stating their preference for imported technologies. On the other hand, discussions with the government research institutes reveal they are not fully informed about the investment preferences of the firms. There is clearly an information gap between the two. In short, most of the Indian venture capital companies being offshoots of existing development banks, despite their autonomous nature, run the risk of exhibiting a similar attitude when it comes to funding risky projects. The prevailing view among Indian venture capitalists is that technology development financing is only a part of venture capital financing and the latter may include other projects as well (Ramesh and Gupta, 1995).

Lessons for Other Developing Countries

In this study, I was primarily concerned with an examination of the institutional support that is required for an effective development of locally generated technologies irrespective of whether these technologies are developed by the government laboratories or by private firms. I identified three principal support mechanisms, namely a correct external environment, availability of risk capital, and finally an institutional support that can correct for crucial information failures between generators of technology and its developers. The second component can in fact encompass the last one. Therefore, of the three the one which is most tangible and the one which can be operationalized is an institutional arrangement for making available risk capital. There is an explicit recognition of this fact and this manifested itself in the form of the government establishing a venture capital fund industry[9]. The US model is the ideal one while in the UK the venture capital industry is supposed to be having a bias against investment in high technology companies (Murray and Lot, 1995).

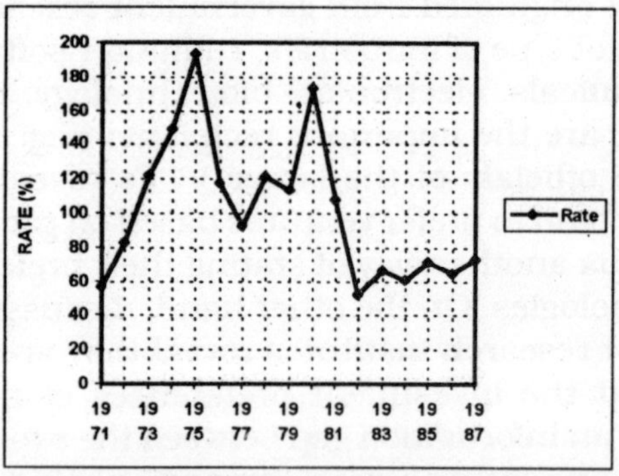

Fig.1. Trends in domestic Technology Generating Efforts in India's Private Sector industry

Much of the support mechanisms that exists in India is primarily geared towards local generation of technology rather than towards its development. Our analysis of the actual operations of the Indian venture capital industry

9. In addition to the venture capital firms there are a number of financial schemes which are primarily operated by leading development banks in the country. Examples of this would be the TDF, refinance of industrial loans, the bills rediscounting scheme, the seed capital assistance scheme, etc., of the IDBI, the SPREAD scheme of the ICICI and so on. We did not consider any of these schemes as an institutional support as they are very limited in their scope and activities.

actual operations of the Indian venture capital industry showed that though they initially started off with precisely the objective of providing an effective financial and information support for domestic technology development, they are increasingly moving away from it. The Indian experience highlights the following lessons:

(i) The venture capital industry is to be regulated through

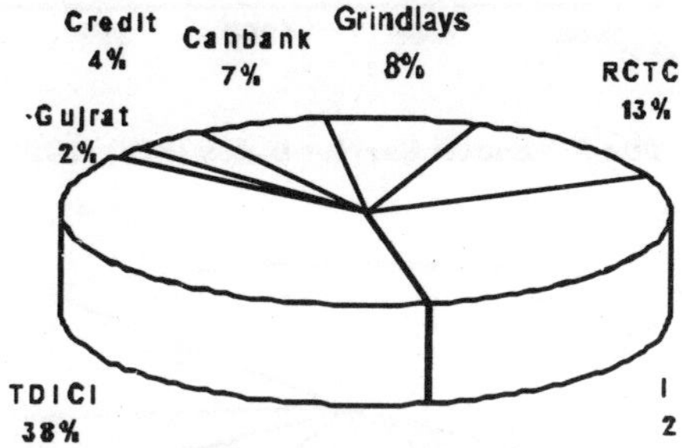

Fig. 2. Firm-wise distribution of venture capital assistance, cumulative 1987-1992

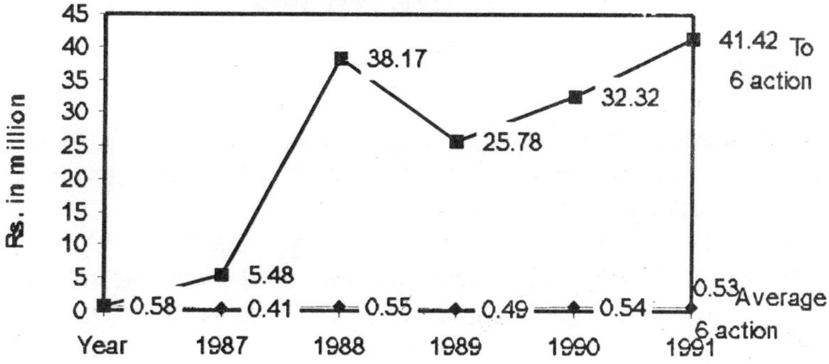

Fig. 3. Trends in venture capital assistance, 1987-1992

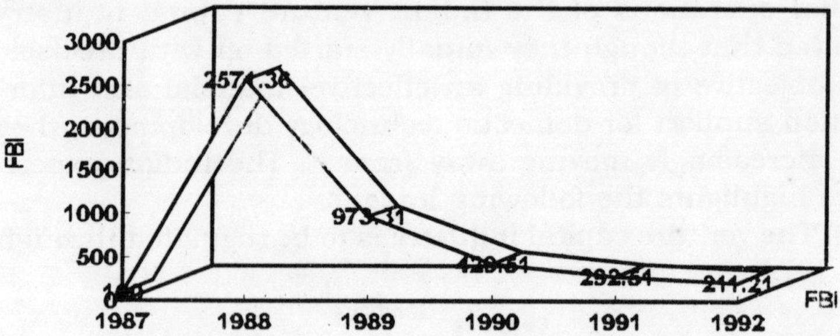

Fig. 4. The Financial Barrier Index (FBI), 1987-1992

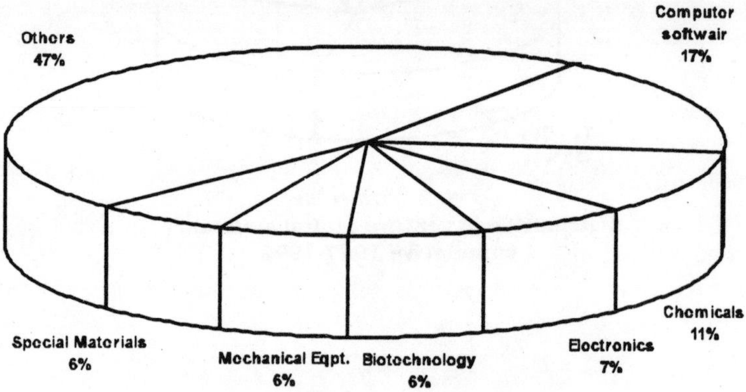

Fig. 5. Industry-wise distribution of venture capital assistance, 1987-1992

government policies;

(ii) These policies should touch upon all aspects of financing by venture capitalists like the stage, type of projects and instruments;

(iii) As far as possible the venture capitalist firms should not be established within the conventional banking establishment as the American experience shows that "credit analysts and lending officers steeped in a conservative financial tradition cannot overnight emerge as venture capitalists". In fact, the moves by Indian venture capitalists to dilute the so-called restrictive provisions of the guidelines

shoots of the existing development banks in the country; and

(iv) In order to make entry into the venture capital industry of genuine players, the tax and other fiscal incentives must be tied to a performance criteria governing their investment operations.

Thus, our analysis shows that the venture capital industry can be an effective solution to the financial barrier to innovations not only in developed countries but also in developing countries, provided the governments have a clearly articulated and enforceable set of rules governing their operation.

REFERENCES

Asian Development Bank (1988). Venture Capital Financing in the Asia-Pacific Region, Manila.

Barabaschi, S. (1992). Managing growth of technical information, in Nathan Rosenberg, Ralph Landau, and David Mowery (eds.), *Technology and the Wealth of Nations,* Stanford University Press, Stanford, pp. 407-434.

Bell, R.M. (1993). Integrating R&D with Industrial Production and Technical Change: Strengthening Linkages and Changing Structures, Economic and Social Commission for Western Asia. UN, E/ESCWA/NR/1993/WG.2/16.

Bhatt, V.V. (1980). Financial Institutions and Technology Policy, *Econ. Pol. Wkly,* May, pp. M-43 M-52.

Boskin, Michael, J., and Lau, Lawrence, J., (1992). Capital, Technology, and Economic Growth, Nathan Rosenberg, R.Landau, and D. Mowery (eds.) *Technology and the Wealth of Nations,* Stanford University Press, Stanford, pp. 17-55.

British Venture Capital Association, (1994). Report on Investment Activity, London.

Brooks, H. (1986). National Science Policy and Technological Innovation, in Ralph Landau and Nathan Rosenberg (eds.), *The Positive Sum Strategy, Harnessing Technology for Economic Growth,* National Academy Press, Washington, DC pp.119-167.

Charles, D. and Howells, J. (1992). Technology Transfer in Europe, Public and Private Networks, Bellhaven Press, London, pp. 59-82.

Cohen W.M. and Levinthal D.A. (1989). Innovation and Learning: The Two Faces of R&D, *Econ. J.*, pp. 99, 569-96.

Cooper, Charles (1994). Relevance of Innovation Studies to Developing Countries, in Cooper, Charles (eds.), *Technology and Innovation in the International Economy*, Edward Elgar and United Nations University Press, Aldershot

Coopers and Lybrand (1994). Fourth Annual Impact of Venture Capital Study, National Venture Capital Association, Arlington.

CSIR (1993). Creating an Enabling Environment for Commercialization of CSIR Knowlędge Base: A New Perspective, A Report submitted to DGCSIR, Pune (mimeos).

Dunning, J.H. (1992). Multinational Enterprises and the globalization of innovatory capacity, Ove Granstrand and Soren Sjolander (1992), *Technology Management and International Business*, John Wiley and Sons, Chichester.

Elizabeth House(University of Oxford) Development Studies Working Papers, N.78, October.

Folster, S. (1988). The Incentive Subsidy for Government Support for Private R&D, *Res. Pol.*, pp. 17,105-112.

Florida, Richard L. and Kenney, Martin (1988). Venture Capital-financed Innovation and Technological Change in the USA, *Res. Pol.*, pp. 17, 119-137.

Government of India (various issues): R&D Statistics, Department of Science and Technology, New Delhi.

Government of India (1988). Venture Capital Guidelines, Office of the Controller of Capital Issues, Department of Economic Affairs, Ministry of Finance, New Delhi.

Green, Milford (1991), *Venture Capital International Comparisons* Routledge, London.

Guha, Ritu (1995). Selling High-tech Nirvana to the World, The *Economic Times*, July 16, p.15.

Ibanez, Fernan (1989). Venture Capital and Entrepreneurial Development Policy, Planning and Research Working Paper. WPS 53, The World Bank, Washington.

Jethanandani, Kishore (1995). Venture Capital: New Directions, *The Economic Times*, 4 August, p. 8.

Kamien, M.I. and Nancy L. Schwartz (1982). *Market Structure and Innovation* Cambridge University Press, Cambridge.

KPMG (1994). Summary of the Tax Treatment of Research and Development Expenditure in Twenty Countries, London.

Lall, Sanjaya. (1993a). Promoting Technology Development: The Role of Technology Transfer and Indigenous Effort, *Third World Quart.*, 14, (1), 95-108.

Lall, Sanjaya (1993b). Understanding Technology Development, *Develop. Change,* 24, pp.719-53. CSIR (1993). Creating an Enabling Environment for Commercialization of CSIR Knowledge Base: A New Perspective. A report submitted to DGCSIR, Pune (mimeo).

Lall, Sanjaya and others (1993c). Bank Lending for Industrial Technology Development, Operations Evaluation Department, World Bank, Washington, Report No: 12138, Volumes 2.

Lall, Sanjaya (1994). Technological Capabilities, in Salomon,J.J., F.R., Sachs-Jeantet, C. (eds.), *The Uncertain Quest, United Nations University Press, Tokyo,* pp. 264-301.

Mani, Sunil (1992). Foreign Technology in Public Enterprises, Oxford and IBH, New Delhi.

Mani, Sunil (1994). Financing Technology-Based Small Firms and Venture Capital Firms: The Indian Experience, The Centro Studi Luca d'Agliano Queen.

Mansfield, E. (1985). Pubic policy toward industry Innovation : An International Study of Direct Tax Incentives for Research and Development, in Clark,K.B., Hayes, RH, and C.Lorenz, The Uneasy Alliance, Harvard Business School Press, Mass. pp. 383-407.

Mashekar, R.A. (1995). Matching industry's needs is CSIR's top priority, *The Hindu,* 10 August, p. 25.

Mayer, C (1992). The Financing of Innovation in Bowen, Alex and Martin Ricketts (ed). *Stimulating Innovation in Industry,* Kogan Page, London, pp. 97-115.

Mishra, A K. (1996). *Venture Capital Financing in India,* Shipra Publications, Delhi.

Mowery, David and Nathan Rosenhery, (1993 National Innovation System in Richard Nelson (ed), *National Innovation System,* Oxford University Press, New York pp. 48-50.

Murray, G. C. And J. Lott (1995) Have Venture Capital Firms: A Bias against Investment in High Technology Companies ?, *Res. Pol.,* pp. 24, 283- 99.

National Science Board (1993). *Science and Engineering Indicators 1993,* U.S. Government Printing Office, Washington, DC. (NSB 93-1).

National Science Foundation (1991). International Science and Technology Data Update, Washington, DC., (NSF91-309). National Venture Capital Association, 1992 and 1994, Annual Report, Arlington.

Nelson, RR (1984). Government and Technical Progress, Pergamon Press, New York.

Nelson,RR (1988). Institutions supporting Technological

Change in the United States, in Dosi Giovanni (ed.), *Technical Change and Economic Theory*, Pinter Publishers, London, pp. 312-29.

Nelson, R.R. (1993). National Innovation Systems Oxford University Press, New York.

Odagiri, H.(1992). Growth through Competition and Competition through Growth, Oxford University Press, Oxford.

OECD (1993). Basic Science and Technology Statistics, Paris.

Patel, Parimal and Pavitt, K. (1995). Patterns of Technological Activity, their Measurement and Interpretation in Stoneman, Paul (ed)., *Handbook of Economics of Innovation and Technological Change*, Blackwell, Oxford, pp. 14-51.

Pavitt, K. (1976) Government Policies Towards Innovation:A Review of Empirical Findings, *Omega*, pp. 4, (5), 539-57.

Piatier, A. (1984). Barriers to Innovation, Frances Pinter, London.

Ramesh, S. and Gupta, Arun (1995). *Venture Capital and the Indian Financial Sector*, Oxford University Press, Delhi.

Rosenberg, N. *et al.* (eds.) *Technology and the Wealth of Nations*, Stanford University Press, Stanford.

Sagali, S. B. and Guidotfi, G. (1992). Venture Capital; the Lessons from the Developed World for the Developing Countries, Financial Markets, Institutions and Instruments. pp. 1, (2), 1-59.

Santarelli Enrico (1995). *Finance and Technological Change, Theory and Evidence*, Macmillan. Press, Basingstoke, pp. 3-38.

Stoneman, Paul (1987). *Economic Analysis of Technology Policy*, Clarendon Press, Oxford.

Stoneman, Paul (1993). Government Policy and Innovation, Innovation Seminar Papers, The Economic and Social Research Council, Swindon. UNIDO (1991). Venture Capital Financing, in *Industry and Development*, Vienna.

Venture Economics (1988).Venture Capital in India, a feasibility study, Canada.(mimeo).

Venture Economics (1991). Venture Capital and Technology, New Jersey.

Venture Economics (1993). Annual Statistics, Boston.

Walker, W. (1993). National Innovation Systems : Britain, in Nelson, R.R. National Innovative Systems, Oxford University Press, pp. 159-91.

11

Aspects of Technology Management at the Industrial Enterprise Level

F.M. Machado

This study presents the experience of UNIDO's project entitled "Technology Management and Industrial Informatics for the Industrial Modernization of Costa Rica", that was implemented from February, 1990 to September, 1993.

Dedicated to building an internal capability in individual industrial enterprises for managing technological innovations, besides creating a Center for Technology Management and Industrial Informatics (CEGESTI), with the objective of assisting small and medium industrial enterprises (SME) with training, information and consulting services in technology management (TM), the project was also called upon to assist TM needs of SMEs in other Central American countries.

Therefore, the experience described henceforth refers not only to Costa Rican industrial SMEs, but also to other Central American countries' SMEs.

The typical barriers encountered for the introduction and adequate application of TM concepts and methodologies in those industrial companies are described and specific unorthodox solutions suggested.

In particular, the demonstration project of technology management nuclei in itself proposes a revision of the basic assumptions behind the traditional technical assistance schemes, by recognizing that the identification of critical technological needs of industrial enterprises is not a trivial matter, that it is rather tied up to their competitive strategies and that their true satisfaction usually requires not shelf-solutions, but very creative ones.

Furthermore, it emphasizes the active role that must be played by the industrial companies themselves in diagnosing their technological needs, using external technical inputs within a creative broader context, assimilating them and starting a new cycle all over in a never-ending process, to be able to continually sustain their competitiveness.

Barriers for Technological Management at Plant Level and Possible Solutions

The adopted macroeconomic policies on liberalization of trade, retrenchment of the State's economic role and other economic structural adjustment policies, coupled with the globalization of commerce, high rate of technological innovations and economic downturns everywhere, are presenting small and medium industrial companies (SMEs) with a much stiffer competition and new challenges which threaten their own survival in great numbers.

Maintenance of standard and quality of products by the firms is more important for surviving in the business rather than a source of differentiation.

In the medium and long run, the firm's competitiveness is based on her ability to learn continuously and to build— at a lower cost and more rapidly than her competitors — the core capabilities that enable her to generate new products, services and other dynamic competitive advantages which her competitors cannot easily imitate or even foresee.

Therefore, the adequate management of those core capabilities and resulting innovation processes — what is called technology management (TM) — can be singled out by itself as the firm's central core capability.

On the other hand, the systematic identification of a firm's critical technological needs for her sustained competitiveness is not a trivial matter. It requires the application of specific diagnosis and auditing methodologies application of specific diagnosis and auditing methodologies against the background of clearly defined strategies for the firm's future development.

All the above leads to a conceptualization of a company's modernization or restructuring process as "planning, timely implementation, control and evaluation of a congruent and integrated set of significant changes, both inside the organization and in its immediate environment, at the technological, organizational, cultural, managerial, strategic, financial, infrastructural and other levels, in response to radical changes in her environment and/or internal conditions, in order to fulfil its competitiveness, growth rate and profitability objectives".

Therefore, this modernization or restructuring process represents a peak segment of a process of change that should have been continuous in the past, but which remained dormant by not only the comfortable refuges created by import tariffs, other nontariff barriers and economic recession, but also by a lack of managerial capability on strategic management and management of change, particularly the management of technical change, or technology management.

By accepting it as such and by analyzing the above definition of an existing industrial enterprise's modernization process, some conclusions can be advanced:

- There is no such thing as a standard competitive profile or "best practices" for an enterprise of a given sub-sector, based on a checklist of factors such as quality control procedures or circles, type of organizational structure, application of just-in-time techniques, marketing activities and others, as assumed by sub-sector diagnostic surveys.

The certification for the ISO 9000 technical norms, for instance only provides the company with one of the conditions required to take part in the European market-place (for the moment). It does not ensure a single cent in sales, which will not materialize unless the company successfully defines and implements her own creative competitive strategies.

Each enterprise's specific set of changes must be oriented

by an equally specific strategy, designed in view of her particular competitive situation and business conditions.

Therefore, it is suggested that following recipes designed by experts for all subsector enterprises not only do not guarantee competitiveness, but can actually threaten the enterprise's own survival.

To start with, strategies based on imitation are transparent to competitors who have developed and mastered them, rendering the imitator defenceless.

Also, by fostering imitation, the required attention from the enterprise's need to develop its own competitive strategy in a continuous way is deviated (when properly formulated).

Finally, those recipes almost always represent the strengths of the industry's leaders, and it is usually plain suicide to try to compete with entrenched leaders using the weapons developed by themselves some time ago. The globalization of markets is increasingly turning this competition unavoidable; there are very few safe havens left.

Thus, if diagnostic surveys are the only approach to industrial modernization, many Central American managers might feel trapped in a seemingly endless game of catch-up-only to be regularly surprised by the new accomplishments of their new future competitors, brought about by the liberalization of trade.

Accordingly, any valid benchmarking represents only a set of reference parameters for strategic definitions as to the innovation processes to be undertaken.

Therefore, the conclusion is very clear: there is no substitute for industrial modernization at the enterprise level, and a country's industrial modernization is the sum of individual enterprises — and their cluster's — modernization, structural competitiveness factors notwithstanding.

- On the other hand, the enterprise's technological decisions and actions in response to macroeconomic policies, including liberalization of trade, should be taken within the framework of her modernization programme and are concentrated on specific corresponding technological changes and their relation to other organizational, cultural, managerial strategic and infrastructure changes.

In accepting the above, how can an enterprise take technological decisions and actions if it has no orientation (strategy) as to the what, why, when, how, who, how much, and where of her technological changes?

Strategy-making is the channel through which the enterprise perceives and analyses its environment and reacts in anticipation of the effect of macroeconomic policies change on this environment.

If there is no strategy, the enterprise will only present some reaction to policies like the liberalization of trade once its full effects are felt and it would usually be too late.

The increased competitive pressure created by trade liberalization might elicit a technological component in such reaction, but even technological innovation by technology transfer takes its time.

In other words, effective technological solutions for increased and sustained competitiveness necessarily require time, foresight and a strategic framework.

Therefore, technological planning must be placed within the overall strategic planning framework of the enterprise, including marketing, financial, production and human resources aspects, among others, which conform to the enterprise's technological decisions and at the same time are conformed to by them, in an interactive process.

Through this exercise all the dimensions of the enterprise's technological decisions (what, when, how, etc.) are made explicit and a convergent collective effort may be developed to effectively implement them.

Thus, it is possible to clearly identify, among others:

- the relative importance of technologies for the enterprises's competitive strategies of product or service differentiation, cost differentiation or diversification routes and by consequence, which technologies are key and which require only monitoring;
- if technological leadership is desirable, achievable and sustainable and if not, which alternative strategy (defensive, opportunistic, etc.) is the most indicated;
- the appropriate origin of required technologies — either through licensing or technology transfer in

general, including reverse engineering, or through in-house or outside research and development (R&D), or a blend of both R&D and technology transfer;

- when technological changes have to take place, through the analysis of product and technologies' lifecycle stages by prospective studies and other mechanisms;
- how to rapidly implement technological changes in harmony with simultaneous organizational, cultural, managerial and other changes in order to avoid "hidden costs" and accelerate new product development time;
- Who are the people responsible for planning, organizing, implementing and evaluating the success and failure of technological change projects?

All the above are among the concerns of the technology management field of expertize at the enterprise level, which has proven to be a helpful instrument in defining explicit overall strategies and catalyzing the set of changes required by SME's modernization processes.

It is important to highlight that, contrary to what is assumed by the existing economical and industrial policies, the large majority of companies in the Latin American industrial sector—in particular small and medium sized companies—do not have the internal capacity to design and implement the change set which represents their modernization process of technological changes mainly-oriented by an appropriate innovative competitive strategy.

There are several reasons for this:

- The majority of the manufacturing companies are traditionally family owned, which in general implies an aversion to risk taking and a necessity for short-term profitability, to ensure the financial survival of the family. This is conducive to a lack of long-term vision regarding investments in important change processes, and a negative attitude concerning the search for strategic alliances, as these may result in a loss of total control over the company and its policies.
- As mentioned before, SMEs do not have strategic capability. As a result, they neither have sufficient

knowledge about the characteristics of the markets in which they operate, including their competitors, nor about their own critical technological competitive needs (they also lack methodologies and information required). In the same token, they do not have the capacity to manage, change or to innovate technological management and are not interested in enabling their human resources or investing in information and knowledge.

Therefore, they are prisoners of their own mental paradigm— the assumptions, beliefs and implicit and explicit values — "the way we do things around here" — which have given positive results in the past but are dysfunctional in the new competitive paradigm.

Thus, there are no ready-made prescriptions for a firm's competitiveness, and the current modernization of SMEs requires a movement in the opposite direction of their natural instinctive movement, derived from the result of their past experience and developed within the import substitution economic model.

The solutions given to problems in the past are not only valid any longer but have transformed themselves into the seeds of the firm's current problems.

In fact, the metaphor of the Delta glider or "handglider" adequately illustrates the situation. When flying in a Delta glider which is in a downturn fall, an inexperienced pilot usually acts instinctively and throws himself or herself backwards, thus increasing his or her fall. The right movement would be counterinstinctive: leaning forward which will allow the pilot to gain height.

When the effects of the economical adjustment programmes, including tariff reduction, are still not fully felt, companies that are already encountering difficulties are acting instinctively — leaning backwards in their companies handgliders — and adopting measures such as:

- more control and short-termism,
- generalized cost cutting,
- raw materials substitution (lower quality),
- indiscriminate firing of personnel (higher costs first),
- reduction of product lines and product variety,

- cancelling new investments, risk - taking and training programmes,
- centrallizing decision-making processes,
- adoption of autocratic leadership styles as the norm,
- exclusive focus on existing clients,
- nonrelated diversifications (into commerce, agriculture, etc.),
- imitation of successful firms' strategies,
- lobbying for governments subsidized funds for equipment renewal,
- others in line with the short-term vision adopted, including an obsessive concern about the short-term profitability of their modernization process.

All the above lead to typical situations in which there is a high level of insecurity, distrust, loss of valuable resources and quality, besides other barriers to the development of the creativity, knowledge and skill, commitment to objectives and innovation processes required to increase and sustain the company's competitiveness.

On the other hand, the counterinstinctive movement required by their restructuring process—leaning forward on the companies handglider—should be characterized by:

- the development of a strategic value adding vision, which should orient not only cost towards reduction processes, but also sales increase programmes, not necessarily the same;
- the design and implementation of creative and dynamic competitive strategies, characterized by multi-layered competitive advantages and by breaking the mould of existing mental paradigms;
- the development of critical core competences for the firm's sustained competitiveness, emphasizing technology management skills;
- identification of the integrated set of innovations required by the competitive strategies adopted and carrying out the corresponding investments;
- placing-value on participative leadership styles, decision making decentralization, human resources skills, knowledge, creativity, motivation and commitment;

- the abdication of the firm's total control, in line with requirements stemming from strategic alliances established with other organizations such as suppliers, clients, universities, R & D centres, competitors and others, with a view to increasing and sustaining the company's competitiveness in globalized markets.

It must be stressed that the above does not imply demeaning the entrepreneurs' fully recognized intelligence and skills. The described predicament stems mostly from the natural and traditional entrepreneurs' dedication to action, as opposed to systematic analysis. Time is SMEs' most prized resource. In addition, the typical accumulation of functions found in SMEs leaves almost no time for systematic analysis and planning. Furthermore, technology management is a new field with ever-evolving frontiers, and information sources are not readily available, among other difficulties.

In spite of those reasons, it remains clear that the modernization process of industrial SMEs of the region, required by their new macroeconomic context, demands substantive and urgent external help.

For the reasons exposed, SMEs by themselves will not be able to tackle the set of radical changes required from them and their mortality rate would certainly bring about very serious economical and social consequences.

How to break this vicious circle? What kind of institutions are the most capable ones for providing the external assistance needed by SMEs which can be saved?

Given SMEs proven resistance to outside training programmes, lack of slack resources for tackling matters not related to immediate survival and so on, the following approach might be a possible solution.

Firstly, the preparation and ample diffusion of audio-visual materials, manuals and computer software dedicated to breaking the old mental paradigm mould and to allow SMEs to carry out in a private way the self-diagnosis of their own problem, should be carried-out.

A simultaneous massive design and implementation of TM training programmes by universities (particularly MBA programmes), TM centres and industrial consulting enterprises should also represent a valuable instrument in reaching some entrepreneurs and adequately forming future ones.

Finally, demonstration projects for creating an internal TM capability in existing SMEs and the development and strengthening of small and medium industrial consulting enterprises should be implemented, together with the complementary financing schemes required.

The following sections of this document describe a successful TM demonstration project carried out in Costa Rica.

Project Background and Set-Up

Even before the publication, in 1990, of Michael Porter's landmark work "The Competitive Advantage of Nations", Costa Rica's Ministry of Science and Technology had launched the aforementioned pioneer demonstration project, based on the conviction that the country's competitiveness depended on the ability of her industrial companies to innovate continuously, rather than on static comparative advantages such as the cost of labour, availability of natural resources, interest rates and other traditional factors.

The progressive branches of metal works, chemicals, pharmaceuticals and computer software were singled out, and 34 individual SMEs were surveyed during a preparatory assistance period, to identify prospective project participants.

The selection criteria were based on aspects identified in the survey, such as the company's general competitive strategy, the nature of its specific competitive environment challenges, inventory of inventions and innovations carried out, strategic and technological planning activities developed, characteristics of the existing organizational culture and structure, profile of directors, technical capability of professional staff and last but not least, the willingness to commit its share of the required resources for taking part in the project.

A batch of ten companies were selected and specific cooperation agreements signed with each of them.

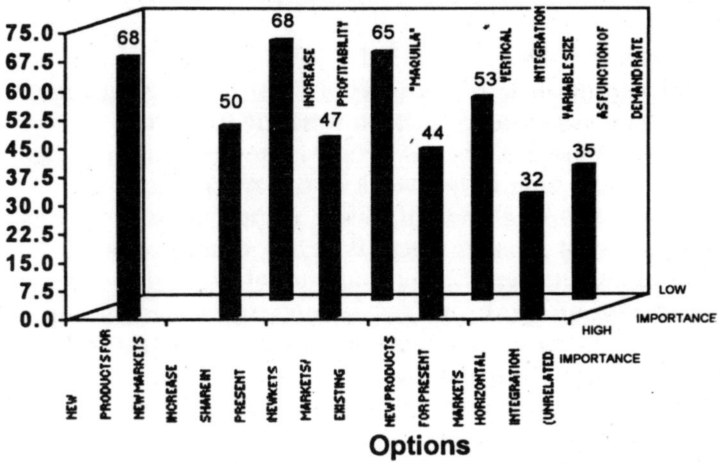

Fig. 1. SMS's General Strategy Options of SMEs for Competion

In view of:

(1) the original intention of establishing a self-supported capability for TM within participating SMEs in accordance with the justification presented in the previous section;

(2) the need to develop continuously technological innovation projects stimulated both from strategy and spontaneous creativity exercised by all people within the organization, connected to opportunities and information generated by chance and environmental discontinuities;

(3) the nature of the general competitive strategies favoured by the companies in the sample, that is, new products for new markets and the inherent confidentiality requirements involved, all characterizing the need of a fulltime TM function, the project design was oriented towards establishing the so-called technology management nuclei (TMN) in each of the selected companies.

These nuclei were created at the highest level of decision-making, with the aim to improve and sustain the firm's competitiveness through technological innovations directed at productivity increase, quality improvement, development of new products and processes and other organizational changes. A TMN consisted of a team of two professionals — a TM expert (the nucleus coordinator) and an expert in industrial informatics — both working full-time, and joined, as it timed-out, by several people in key positions dedicating part of their time to the work of the nucleus.

Cooperation agreements were signed for a period of one year, the cost of each nucleus being shared on a 50-50 per cent basis between the company involved and the project. A relatively modest budget—covering salaries, specific technical assistance, access to databases, technical journals subscriptions, acquisition of patent and technical norms documents, computer hardware and software and study tours to technical fairs or visits to technology suppliers, besides other typical operational expenses — enabled each nucleus to introduce effectively TM concepts and methodologies to the company, and served as a springboard for the formulation of technology innovation projects, to be financed by the company itself or through external sources, such as science and technology government funds, banks and so on.

On the other end, the simultaneous creation of CEGESTI enabled TMNs to seize available external technical assistance services in TM training, information and consulting.

Among others, each cooperation agreement signed included provisions with respect to the nucleus organizational positioning (staff to the Managing Director), the results to be produced by it (at least 5 technological innovation projects formulated yearly — of which two should contemplate the introduction of information technologies and one should be of a pre-competitive nature, of interest for the specific sub-sector), besides a preliminary work programme, budget and evaluation and control mechanisms (the steering committees were formed by the Managing Director, a representative from the Ministry and the UNDP - UNIDO project director).

Therefore, each TM nucleus worked as a beneficial technological "Trojan Horse", creating a slot of organizational space and resources to deal not with the company's day-today routine operation, but with the measures required to ensure sustained competitiveness, with emphasis on technological innovations.

Through their establishment, SMEs were provided with the required protagonists to balance present and future survival and adequately deal with confidential information not easily trusted to outside consultants.

Furthermore, in fulfilling its role, it was expected TMN's staff would also create and maintain the required links with universities, research and development (R&D) centres, suppliers, clients, R&D funds and other agents of the technological innovation chain, besides becoming the internal champions for innovation and change in general, to the benefit of each company's modernization process.

By catalyzing strategic management and linking it to technology, TMNs were expected to identify critical technological needs for the company's sustained competitiveness, generate a corresponding portfolio of explicitly formulated technological innovation projects (R&D, technology transfer and other types) within clearly defined technological strategies, help find outside financing and effectively implement and manage them, "in tandem" with other financial, organizational, managerial, and cultural innovations.

In doing so, TMNs were also expected to be instrumental in providing adequate answers to, among others, the following questions:

- How to identify the critical technological needs for the company's sustained competitiveness and the business profitability? How to obtain the corresponding technologies and put them to use in an effective way?
- How to design a successful combination of technologies/products-services/markets?
- How to select suppliers of innovative technologies? When and how should the company transfer those technologies?
- How to generate more competitiveness, growth and profits through the company's actual and future technological assets?

- Which are the key factors determining the accelerated development of new products and services? How should they be managed?
- What new competitive strategies can be generated by those technologies?
- How to compare the company's level of technology utilization with that of her competitors?
- How could the company's leadership effectively promote innovation?
- How to improve the quality and productivity of the company on a continuous basis?
- How to successfully create and develop "spin-offs" or new technology-based companies?

In short, TMNs were considered as a pre-investment mechanism necessary to:

- manage technical and market information in a systematic way to feed the process of defining specific competitive strategies;
- identify SMEs critical technological needs and opportunities explicitly linked to the company's (and its production chains) competitiveness, growth and profit generation objectives (not trivial); formulate technological innovation projects, submit them to outside financial sources and negotiate their approval;
- manage those projects in such a way as to ensure the effective participation of all involved and the efficient achievement of specified objectives, whereby providing financial sources with additional assurances as to project success possibilities and consolidating corresponding strategic alliances with suppliers, clients, universities and so on. Therefore, through those activities, TMNs were expected also to open up the way to strategic alliances with other organizations and to provide the missing link for the linkage industry/research institution, since — as the saying goes — "it takes two to tango".

Project Results Evaluation and some Lessons Learned

Apart from UNIDO's backstopping, the project underwent three different evaluation processes—two outside

Development Programme UNDP— on the soundness of its concept and on the TMN results, besides one final appraisal of TMNs carried out by CEGESTI and presented below.

All of them gave more encouraging results than initially expected, given the pioneer nature of the project.

It is important to note that one agreement was terminated prematurely, due to changes in the company's priorities and a second was established after the first two batches of cooperation agreements were implemented, the reason why the final evaluation (Tables 1 & 2) is restricted to seven companies.

It is also worth mentioning that one TMN was established in a state company and another in an association for metalwork micro enterprises, as further experiments on the possible breadth of the TMN concept application.

Tables 2 to 5 show the evaluation of both the direct and indirect impact of the TMNs.

Table 1
Participating Enterprises in the Final Evaluation

Name of the Company	*Number of Employees*	*Branch*
Aptamai Association of Owners Industrial Maintenance and Repair metal work Shops	6 (headquarters)	Metal-processing car maintenance and repair, painting and solding
Celco Celco of Costa Rica	65	Chemiclas
Nexys/Dedisa	35	Informatics-Software
Gutis Gutis Products inc.	70	Phamaceutical products
Ice Costa Rican Institute of Electricity	8,000	Power generation and telecommunication
Seyma	41	Safety,

Maintenance Inc: **Tecapro** Tecapro of costa Rica Inc	45	Security and indstrial maintenance Informatics-Software

Table 2
Final Evaluation-Summary of the TMN's Direct Impact

Direct indicators of the tmn impact	*Results (percentages)*	*Observations*
1. Development of new production processes	100 per of the enterprise	All enterprises indicated sign-ificant or very sigificant im-pact of the nucleus
2. Development of new applications for existing products and/or services	67 per of the responding enterprises	All respondents signalled a significant or very significant impact of the nucleus
3. More flexible production	83 per cent responding enterprises	83 per of the registered case indicated a very significant impact of the nucleus
4. Optimization product variety and available resources	67 per cent responding adjusted	67 of the respondent indicated a very significant impact of the nucleus.
5. Quality improvement in products and/or services	100 per cent responding firms upgraded	All enterprise registered a very significant impact of the line nucleus
6. Performance	100 per cent of	In all firms the

6.	Performance Improvements in products and/or services	100 per cent of responding firms improved their product perfor- mance	In all firms the impact of the nucleus was rated as signi- ficant or very significant
7.	Improvement in products and/or services	67 per cent of the responding firms improved their product perfor mance	In all firms the impact of the nucleus was rated as signi- ficant or very significant
8.	Image improvements products and/ or services	83 per cent of repondents	Attributed to the efforts of the nucleus
9.	Quality improvement in sales services	100 per cent responding firms	Attributed to efforts of the nucleus

The results speak for themselves. It is nevertheless important to highlight the 112 projects generated and formulated by the 7 TMNs — well beyond the number expected.

As a result, the majority of the technological innovation projects financed by the most important science and technology fund of Costa Rica, managed by CONICIT — the National Council for Scientific and Technological Research — during 1991 and 1992, were projects presented by SMEs participating in the project.

This fact demonstrates that TMNs worked as effective instruments for making technological demand from SMEs explicit.

Greater links with R & D centres and universities were also detected.

As expected, TMNs promoted strategic management, organized and implemented technical and market information centres within individual companies and functioned as innovation champions, using technological innovation projects as catalyzers for other types of innovations such as managerial, organizational, cultural and strategic innovations, among others required by the comprehensive and integrated modernization process of the enterprises involved.

Some Lessons Learned

The following aspects were considered as critical for the TMNs' success:

1. A previous and clear understanding of the intrinsic nature of TMN activities and the scope of projected results of the nucleus by the SME's board of directors and senior personnel.

Given the novelty of TM concepts and methodologies — usually not within the current mental paradigm of SMEs — and the fine threshold line dividing preservation of the present and building of the future, companies showed a tendency for dedicating TMN's priorities to short-term operational troubleshooting. Although this is unavoidable to a certain extent, in order to create internal short-term credibility for the nucleus, a balance with activities of a more strategic nature should be mandatory.

The above was achieved in the project by other measures:

- carrying out a half-day awareness workshop with prospective SMEs' directors and senior personnel, in which basic TM concepts were discussed and technology's role in solving the existing barriers for the achievement of current objectives and goals, in benefiting the firm's position with respect to critical competitive factors and in other aspects of organizational life was made explicit.
- securing full transparency and consensus on obligations, responsibilities and rights of the different parts involved in the cooperation agreements.
- a constant external follow-up of the TMN programme and regular meetings of the steering committee to solve differences in points of view and to assess the achievement of jointly set goals.
- carrying out TM seminars designed to co-opt the full cooperation of senior personnel in key positions in

order to facilitate the nucleus' organizational induction, acceptance, support and protection.
- exercising external "leverage" on the TMN coordinator, through his or her recruitment, selection, appointment, salary payment, constant peer meetings and performance assessments.

2. The SME's managing director's personality, strategic vision and leadership style, which provide the future "window framework" and conform to the organizational culture that stimulates or inhibits risk taking, creativity and innovation, amongst other requirements for TMNs'success.
3. The TMN's ability to bandwagon with the SME's strategic management activities or catalyze them, if nonexistent. The failure to do so practically limits the nucleus role to short-term troubleshooting, undermining the fulfillment of its mission.
4. The quality of the human resources deployed.

The major difficulty concerns finding experienced TM managers. The project resorted to previous (preparatory phase) and simultaneous training of selected professionals most of whom graduated in industrial engineering.

It is believed the adequate development of a TM manager is as time-demanding and complex as the development of a good scientific researcher.

The experience of companies such as Martin Marietta in the USA shows that to transform a freshly graduated Ph. D into a good technology manager who can also provide technological solutions to problems encountered takes about 15 years. Developing countries lack the stock of Ph. Ds to start with, and are pressed to rely on a different and much quicker route, by forming technology managers in a much shorter time (1 to 3 years) and resorting to external inputs to solve specific technological problems.

In particular, the interpersonal competence—communication, empathy, conflict resolution, negotiation and other abilities—of the TMN coordinator, coupled with a flexible leadership style, has shown to be of paramount importance to the success of the nucleus in dealing with the management of change in general — the technological, organizational, cultural and other kinds of changes required by SME's integrated modernization process.

Furthermore, the match of the coordinator's values with those of his or her designated company, his or her capacity to be motivated by the company's goals and objectives within his or her short-term assignment challenge and his or her continuous willingness to learn are also considered very important factors.

5. As a result of the above and the contingent aspects of human resources development in TM within the project scope, the external technical assistance availability, both TM and otherwise, have also proven to be an indispensable complement.

Therefore, the degree of support provided by CEGESTI in TM and by specific outside consultants in contributing with technological inputs to innovation processes were of a much greater magnitude than originally estimated, and some additional resources from other sources had to be summoned.

6. Extra external TM support should be provided for assisting TMNs in formulating explicit technological innovation project documents, with background information, justification, objectives, results, time tables, description of inputs, detailed budget and so forth.

Most SMEs were found to be very resistant to carry it out for a variety of reasons: the fear of having one's idea stolen, not knowing how to do it, paperwork equated to "bureaucracy" and so on.

Given the outweighing benefits of the exercise, short workshops on project planning and control were the most useful instrument to change this mind-set.

7. To positively seize the emulation phenomena. Certain parts of SME's organizations felt threatened by the implementation of the nucleus and started developing typical TM activities in direct competition with it. Appropriate strategies for co-opting those efforts have shown to work as accelerators of the nucleus organizational induction and early contribution to the achievement of SMEs' goals.

8. The time frame adopted-one year— for the cooperation agreements was clearly inadequate. The selection and basic training of appropriate TMN staff, the "in-loco" detailed revision of the original work programme, the procurement and allocation of specific inputs, among others, took a much longer time than originally expected, and the definition and implementation of changes within SMEs also require a certain amount of negotiation and time.

Most of the cooperation agreements were extended in time by request of the SMEs involved, but no additional resources apart from time were made available.

From the project's experience, 1.5 to 2 years is the adequate time frame for consolidating the work of TMNs.

It is expected that this limited but very encouraging experience can be replicated elsewhere — with the necessary contextual adjustments — in order to advance the understanding of the nature of the mechanisms required to contribute towards an increased and sustained competitiveness of industrial SMEs in developing countries.

BIBLIOGRAPHY

"Coping with a changing agenda" - Taylor, R.C. 1992 International Strategic Management Conference, *Planning Rev.* p. 20.

- Fragmentacion y Contexto de la Gestion Tecnologica y su Impacto en la Formacion de Recursos Humanos", Machado F.M. in "Aspectos Conceptuales y Metodologicos de la Gestion Tecnologica", IDRC/ALTEC - Ediciones Dolvia C.A., Noviembre 1992, Venezuela.

"Gestores Tecnologicos o Emprendedores de Origen Tecnico? "Modernizacion Industrial en Centroamerica: "Quid pro quo". — Machado Fernando M. Espacios - Revista Venezolana de Gestion Tecnologica, December 1992, Vol. 13, No. 3, Caracas, Venezuela.

"Managing Strategic Change - Strategy, Culture and Action", Johnson, G. *Long Range Planning,* pp. 25. (1), 1992.

"Nucleos de Gestion Tecnologica jActuar donde impacta: Al interior de la empresa! Rodriguez, "Javier I. Van Notten y L. Saborio, ALTEC Biannual Congress Proceedings - Venezuela, 1991.

"Strategic Intent" - Hamel G. & Prahalad, C.K. *Harvard Business Rev.*, May/June, 1989.

"Viabilidad de Innovaciones Tecnologicas en las Empresas Industriales de Costa Rica" Machado Fernando M. Ana y. L. Retana A. — ALTEC Biannual Congress Proceedings, Brazil, 1988.

12

Success Factors for Market Responsive R&D

Soumyo D. Moitra

The importance of R&D to a firm's competitive strength has now been amply established. Indeed, the importance of technology management and the role of innovative products to a firm's long-term success is taken as axiomatic in today's management discussions. What is still at issue is how best to manage R&D, technology, and innovations.

This is probably one of the key areas for investigation today, because we know far less about the implications of decisions in this area than in other, more established areas such as inventory control or production. Thus the study of the factors that influence the success or failure of projects and innovations is of paramount importance for R&D management. Much research has been done recently to identify these factors, and to find out what managerial decisions help to improve the chances of success. It is with chances that we have to be necessarily be concerned with, since R&D is inherently a risky endeavour, and there will inevitably be a significant number of failures as scientists and researchers explore the frontiers of technology to find out what works and what does not.

This study reviews the current research on the factors that influence the success or failure of R&D projects and innovations in industry. The literature on this topic has been reviewed to see what factors have been either identified or suggested as having an impact on the outcome of R&D

projects. These factors form a very diverse set and cover a wide range of issues. In order to arrive at a clearer overview, the factors are classified into three groups: structural, strategic, and tactical. The structural factors are those that are related to the organizational structure within a firm. The strategic factors are the decisions the firm makes regarding its long-term corporate and technological goals. The factors related to tactical planning are the many operational decisions that are taken with respect to R&D management. These are the detailed decisions that have to be taken regularly, and sometimes very often.

In addition, when we consider the important factors in R&D decision making, we need to keep in mind for what outcomes they are important. For market responsive R&D, it is the production of innovations and their market performance that is most important. Some factors affect the rate of production of innovations, that is, how innovative and R&D oriented the unit is, and some factors which affect the performance of the innovations in the market, that is, what revenues, or profits, or market share it brings in. Some factors may affect both of these outcomes.

To clarify the scope of this study, it should be stated that we are considering industrial R&D only, that is, where commercial success is the major goal, and we envisage R&D to include all the stages from initial ideas to the end product and even initial market introduction. Although this paper focuses on in-house R&D of firms, many of the factors are also relevant for a successful partnership between independent R&D organizations and external firms. Therefore, these same factors would facilitate R&D organizations to become more market responsive. Also, there may be some factors or decisions that do not matter, and it is important to recognize them so that time is not wasted on them at the expense of the significant factors. However, that issue is beyond the scope of this study.

In the next section the classification scheme that is used to group the various factors related to R&D is introduced. In section three, we review the significant factors in each of the groups. In the fourth section we focus on four key factors and discuss some ways of implementing them. In section five we consider the implications of the findings, and possible policy directions. A summary and a discussion of future research that is needed is given next. Finally, a selected bibliography of some relevant books is provided.

Classification of Factors

In reviewing the literature and the factors that the various articles, reports, and books given in the bibliography have identified as significant to R&D results, it became clear that there was a very large variety, and in order to have a systematic approach, it would be necessary to develop a classification scheme. The following three general classes were selected because they seemed the most useful and natural. Also the terms have been widely used in the management literature to classify strategic concepts. The groups are: structural, strategic, and tactical.

The structural factors are those that are related to the organizational structure such as the divisions within a firm, the nature and relationships of the R&D groups, the flexibility of the group structures, the interfaces among the groups, flexibility of group structures, communications channels, the organizational climate, and group dynamics. For a more detailed discussion, the reader is refered to Refs. 1 to 6. The strategic factors are the decisions the firm makes regarding its long-term corporate and technological goals. Examples of these are the choice of technologies and markets, the length of the planning horizon, the kind of R&D processes that are developed, the emphasis on developing technology, how technological planning and goal-setting is done, the reward system, interaction with the environment, and the technological alliances that it forms. These issues are discussed in most of the references, in particular in 7 and 8. The factors related to tactical planning are the many operational decisions that are taken with respect to the allocation of resources, R&D project selection, product design, marketing, and project control. These are the detailed decisions concerning the tasks and subactivities within the planning process that have to be taken regularly, and sometimes very often. A set of key decisions in this group are those related to the implementation of the new product development process (9-11).

The Significant Factors

We now discuss the major factors according to the above classification scheme.

A. Structural Factors :

The following structural factors appear to have an important influence on the success of a firm's R&D efforts.

(i) Geared to Strategy

The firm's structure has to be geared to its strategy and corporate goals. This has been well known in the strategy literature for a long time (7, 8) and research on R&D and innovations has confirmed this (5, 6). When this is absent, a firm might make technological breakthroughs, but will fail to benefit from them because there is no strategy to guide it to do so. Case studies in the literature include those on Xerox Parc (USA) and the former GE-Microelectronics Group (USA).

(ii) Exploit Existing Competencies

The firm must recognize its own competencies and develop an appropriate structure to exploit them. Otherwise it might drift into areas where it does not have a competitive edge, and might not focus on areas where it has. Thus this is an extension of the first point above. Further, there has to be corporate recognition that R&D and innovations play a key role in future profitability.

(iii) Implement Technologies

It is very important for a firm to implement those technologies that support its current business. A firm may have been at the leading edge in some area, but if that area has lost its commercial importance, then clinging to it could be disastrous. The firm needs to implement new, relevant technologies where it can have a business advantage and for this it has to develop the suitable structure. Case studies 1 and 2 again have illustrated this; for example, General Radio (USA), which also changed its name to GENRAD to reflect its new thrust.

(iv) Dynamic Groups

In the area of R&D and new technologies, the mix of skills required for success changes continually. Thus old groups need to give way to new groups on a regular basis. Thus the firm needs to be able to manage dynamic groups effectively.

(v) Synthesis

Increasingly, high technology projects require many different kinds of know-how. That is, the firm must be able to create a synthesis of different skills for successful R&D. Some firm structures facilitate this multifunctional participation while others hinder it. Firm structures that can achieve this help in the success of projects.

(vi) Efficient Interfaces

In addition to the internal structure of groups, all R&D projects require interactions between different groups. For this, the "interface" between groups has to be good, that is, groups communicate with respect and trust each other. Research indicates that the interfaces between R&D (technology), marketing, and production are particularly critical (6).

(vii) Interface for Alliances

When the firm has an alliance of some kind with another firm, (and this is becoming more common because of the scale and complexity of R&D projects), it is best to have a separate group to interface with the partner firm. This appears to be a better solution than have the R&D group that is busy with some project get the additional task of interfacing with the partner. This tends to distract the R&D group from its original mission.

(viii) Top Management Commitment

This issue is actually quite complex. While top level commitment and leadership are undeniably important, senior managers need not be involved with the details of R&D projects. They should however be kept informed from the early stages. What successful projects tend to have are "champions" who get support for the projects from top management, but are not necessarily senior managers themselves. These champions should have control over the specific project. Another role that is helpful to success is that of a gate-keeper. This is a person (or persons) who scans the technological developments outside the firm, and reports relevant information back to R&D personnel within the firm (3).

(ix) Good Communications

This is a well-known requirement, and R&D research, as reported in all the references, also confirms this. Some form of "inter-company networking" should be promoted as company policy.

(x) Collective Orientation (Climate)

This concept is concerned with the culture of the firm, and has several aspects to it. The aspects related to this concept that contribute to R&D success are,—

- common purpose,
- crust,
- teamwork,
- collegality (lack of hierarchy and ranks),
- innovation-oriented mindset.

B. Strategic Factors

As mentioned above, some factors that influence the outcome of R&D projects can be considered as strategic. Current research (see for example refs. 7, 8 and 11) indicates the following as the significant ones. However, there are other factors involved as well, and future research may show that those might be important too.

(i) Technological Choice

The "technology" strategy of the firm is probably the single most important determinant of the success or failure of its R&D effort. In particular, the quality of the decisions related to technological choices is a key factor, although difficult to prescribe. Successful firms consistently make what are described as "sound technological choices", and this is facilitated by including technical people in the senior levels.

(ii) Long Term

One factor that invariably shows up when successful firms and innovations are studied is that there has been a long-term commitment to specific technological goals. The intermediate objectives or tasks may change (and usually does in a rapidly changing technological environment) but the ultimate target is not readily abandoned. In fact, research shows that major innovations have had a surprisingly long development time. An example of this is the Video Casselte Recorder (VCR) where the players (in the USA and Japan) pursued the goal of a video recorder, analogous to the audio tape recorder, no matter how it had to be developed (2) Thus successful firms are willing to wait for a long time before returns to R&D are realized. In the case of the (VCR) it was about 20 years.

(iii) Consistent

This factor is consonant with the long-term objectives. Firms not only have to be patient, but also consistent.

(iv) Learning (R&D Capital; Failures)

It is important for firms to learn and store the knowledge gained from R&D projects, including those projects that might appear to have failed in there objectives. There can also be unanticipated side-products, or later benefits from apparently failed products. The firm has to develop its R&D capital which is embedded in its R&D people.

(v) Clear Planning Process (NPD) — Quality

Having a clear planning process in place helps firms to succeed more frequently with innovations. In the literature this is often referred to as the "New Product Development" (NPD) process (10). In particular, it is important to have effective screenings, reviews, and terminations (when necessary) of R&D projects.

(vi) Choice of Markets — Market orientation

The choice of markets where the firm chooses to be a player is extremely important. This decision has to be driven by a true market orientation which involves understanding segments, customer needs, assessing the attractiveness of markets, and the firm's position in the markets (7).

(vii) Proactive Planning

All R&D and new product/technology planning should be proactive in order to be efficient. Reactive planning necessarily is late and inefficient.

(viii) Environmental Scanning

For a firm to maintain its technological edge, it has to keep up with the developments in the environment. This requires the firm to allocate sufficient resources to "environmental scanning"

(ix) Alliances Selective Trade-offs

Forming alliances has almost become an integral part of R&D and new technology development since no one firm has all the expertize to do the development work all by itself. However, as is well-known, forming alliances is a risky venture, and many case studies have shown the pitfalls that exist. So there is a tradeoff between the advantages and disadvantages, and it is important for a firm to evaluate the advantages it will get. The main set of advantages falls in the class of "complementary assets". That is the partners should possess different but complementary strengths, for example, one firm in one area of technology and the partner firm in another area. Alternatively, the complementary strengths could be in any one of manufacturing, marketing, distribution and services.

(x) Risk Taking

All R&D endeavours involve substantial risk. It is inherently difficult to predict the outcome of experiments and research. Thus firms have to be willing to take risks, and accept the fact that even with very competent staff, there will be some or even many failures. Failures should be taken as a matter of course, and research personnel should never be punished for that.

(xi) Reward System

Successful firms also have equitable reward systems. This is very important because R&D groups may not be associated with the final stages of marketing and product introduction where the contribution of people to the revenues or profits are easier to assess directly. The contribution of R&D to the success of products, though somewhat distant, should be recognized and rewarded.

(xii) Moderate Control of R&D Staff

The control of R&D staff is a difficult balancing act for senior management. The staff has to be given enough freedom so that it can be creative and have job satisfaction. This is essential if the firm is to attract and retain talent. On the other hand there has to be some control so that the R&D efforts focus on the firm's business needs and strategy.

C. *Tactical: (Routine Planning and Opprations)*

In the area of tactical decisions, there are also a number of factors that can influence the probability of success for R&D projects. The following are the major factors hat have been found in the literature (5, 9, 10, 11).

(i) Portfolio of R&D Projects/Innovations

Typically a firm has many R&D projects under consideration. What subset of these projects should be undertaken, that is, the "portfolio of R&D projects" is a very important decision. The criteria for selection, the data for the basis of selection are important, along with the next two factors.

(ii) Process of Selection of R&D Projects

How a firm goes about selecting its R&D projects has been found to be important. The process has to be systematic, as unbiased as possible and should consider at the very least the key characteristics of the R&D projects.

(iii) Set of Factors Considered

The set of project characteristics that are considered in R&D project selection is important for ultimate R&D success, and should include the following:

{Benefits, costs, synergies, complementarities, substitutabilities, technical dependencies, strategic fit, risk, long/short-term completion times, and durability of the benefits}.

(iv) Marketing Competencies

For successful R&D projects, firms have to keep in mind that the end product will have to be successfully marketed as well. Therefore the firm will need marketing competencies and should focus on R&D projects that will lead to products that the marketing department will be able to sell. Potential synergies between end products, and between the R&D department and the marketing department should be exploited to the full.

(v) Market Research

A lot of research (see refs. 10 in particular) has shown that when market research is done upfront, and the results are considered as input into the R&D decisions, the chances of success of the R&D output improve considerably. A market orientation of the R&D unit helps to improve the chances of success, and ideally the R&D groups that developed the product should be involved up to the product launch and sometimes beyond. For radically new innovations, it is important to identify innovative or "lead" customers, and R&D may be able to help in this.

(vi) Feedback from Retailers, Distributors. Sales People, Service Groups

This feedback is a valuable source of information for developing future technologies and products. For example, good rapport between the R&D groups and customers has been found to lead to successful products.

(vii) Competitive Response

In a competitive situation R&D should strive to design products that are close to the "ideal point" in terms of the characteristics that potential customers prefer.

(viii) Product Design — Product Advantage

Another key to successful R&D products is the unique advantages or value that the product offers. This value is again a composite concept that includes quality, performance, cost, features, and unique benefits.

(ix) Allocation of Resources to Key Activities

Research into R&D activities has shown that firms do not allocate resources optimally in order to maximize the chances of success (10). Typically they spend less than they should on the following tasks:

PRELIMINARY ASSESSMENT (marketing and technical)
TESTING (in-house and with customers)

(x) Transfer of Controls

Earlier research appeared to find that when all the tasks in an R&D project are done collaboratively by all the relevant groups (eg. R&D, production, and marketing) then the chances of success are higher. However, more recent findings suggest that while cooperation and good communications are very important, it is not always necessary that all the groups should always be involved. Rather, different groups should be in control at different stages, and control should be transferred from one group to another in the course of the R&D project (6, 9, 10). Each stage should be fully completed before the transfers are carried out.

Focus On Some Key Factors

The factors discussed above are complex, and generally involve subfactors within them. Given in Table 1 of some key factors, and the subfactors that are a part of them. It is important to identify these subfactors because they can provide more specific guidance on implementing the main factors in practice.

Table 1
Some Key Factors and Subfactors for Success in Inventions

1. The NPD process:	Many ideas, quick selection effective screening, mkt/tech, collaboration innovation fully ready
2. Interfaces: Critical	R&D marked, interface transfer of control risk sharing
3. Alliances: (Technology)	Technology clusters, basic-core networks, complimentary assets, degree of match
4. Leadership:	Role of top management, champions reward system

Some solutions to the problems of implementing these factors are:—

* Integrative participation.
* Departments treating each other as business partners.
* Exchanging personnel among departments.
* Forming new groups and/or divisions.
* Forming matrix organizations.
* Providing greater resources/rewards to individuals/groups that produce innovations that are successful in the market.

Implications

A. Policy Directions

This study has identified some major factors that lead to the success of innovations in the market. Although the focus was on in-house R&D, the same factors hold in the case of a partnership between an R&D organization and an independent commercial firm. The R&D organization needs to take cognizance of the factors that make R&D efforts or outputs successful, and those factors that improve the relationship of the R&D group with other groups in the firm will also promote a successful partnership between R&D organizations and industry.

Therefore, in terms of policy directions, the above factors need to be implemented and the quality of the implementation has to be satisfactory. For example, recognitions and reward system for market-oriented R&D achievements may need to be implemented. Then, the reward system has to be implemented in such a way that it is perceived to be fair and adequate.

B. Role of Government

The role of government in promoting market responsive R&D is huge and crucial. Here we focus only on those few aspects that are related to the above discussion.

Promoting awareness of Intellectual Property Rights (IPR) issues and ensuring compliance is clearly important. But beyond that, government should also encourage the development of know-how (within each industry) that will enable R&D organizations and their partner firms to by-pass patents. By-passing in this context means finding legal alternatives to achieving similar goals that a patented technology achieves. Training in the concepts of, and approaches to, reverse engineering is also very important.

Government should encourage exploiting capital-to-labour cost ratios wherever there is a competitive advantage. Such areas may include infrastructural developments that support R&D for a variety of technologies. There may be opportunities to "leap-frog" to the latest technologies, and government can often aid in this.

It would be helpful if government promoted a market-oriented attitude in R&D organizations. For example, it could encourage or assist R&D organizations to search for market niches or spinoffs for its products.

Finally, government organizations could try to identify core technologies needed in each industry and promote R&D to fill any gaps that might be uncovered.

Summary

In this study we have reviewed the literature on R&D management and have identified and discussed the major success and failure factors in R&D. We have classified the factors into three general groups, structural, strategic, and tactical. We have described the major factors in each group, and highlighted some key factors along with their subfactors. Finally, we have discussed the implications of these findings for policy directions and the role of government.

There is clearly much further research to be done. In the Indian context, empirical studies of R&D projects by Indian firms that result in a new technological innovation have to be carried out. More case studies on R&D management have to be developed and the lessons to be derived from them must be articulated. There are currently conflicting evidence and paradoxes in the R&D management literature that call for a resolution. Finally, factors that do not appear to matter very much should be identified and investigated further.

BIBLIOGRAPHY

The strategic Management of Technological Innovation, - Loveridge and Pitt, M. Wiley, Chichester, 1990.

M.L. Tushman, *Readings in the Management of Innovations* - and Moore W.L. (eds), Harper, 1988.

Twiss, B. *Managing Technological Innovation*, - Pitman, London, 1992.

Roberts, E.B. *Generating Technological Innovation*, (ed), Oxford University Press, New York, 1987.

Utterback, J.M., *Mastering the Dynamics of Innovations*, - HBS Press, Boston, 1994.

Souder, W.C. *Managing New Technology Development*, and Crawford, C.M., Mcgraw-Hill, New York, 1994.

Steele, L.W. *Managing technology: The Strategic View*, McGraw-Hill, New York, 1989.

Rothberg, R.R., *Corporate Strategy & Product Innovation*, - Free Press, New York, 1981.

Roussel, P.A., *Third Generation* R&D *et al.* HBS Press, Boston, 1991.

Cooper, R.G. *Winning at New Products*, Kogan Page, London, 1988.

Betz F. *Strategic technology management*, McGraw-Hill, New York, 1993.

13

Advanced Industrial Culture: Advanced Engineering and Information Intensive R&D

Parthasarathi Banerjee

Most of us share a common refrain that in India we lack an advanced industrial culture. Nevertheless we often concede a ground to our advanced industrial achievements, though sporadic. We have failed in integrating such achievements with our other non-spectacular routine life. This appears to be the reason behind our common perception of an absence of the culture of the advanced industry. We also have, however, a common sense that we have to forge forward and in order to achieve that, it is understood, an advanced industrial culture is a necessity.

However, the protagonists of markets and the corporate assert that culture aside, what we require urgently is the strengthening of the corporate and also a market defined through inter-corporate competition. It is claimed by the protagonists that advanced industry comes through corporate innovative adventures. Competition is the key element, they assert. This, however, goes against the evidence. Big corporates have a minority share in the GNP of the OECD countries. Far too little innovations have indeed flowed out of the corporate even in these OECD countries. Moreover the market, as the interrelationship amongst the enterprises, is often only a shared arrangement. In fact, evidence from East, South-East Asian countries as also from India, indicate the pervasive influence of cooperation, as against competition, on the industrialization. Moreover, cooperation seems more at work in the capital-intensive frontier areas of technology than in the "traditional" areas. Further, this cooperation, including that in the advanced areas, appears to be institutionalized around a network (or diaspora). We have evidenced networks in Tirupur, in tiny industries working on silk, and also around those R&D establishments together with manufacturing enterprises who are working in the area of aerospace. We have grounds therefore to doubt 'competition', 'corporate', 'advanced industry through corporate competition' and 'culture as something secondary and traditional'. In order to resolve these doubts, we have studied the network of R&D and related production, around Bangalore, in the area of aerospace. We present a summary of our understanding.

In this article we have defined advanced industrial culture as an attitude of working with "advanced engineering and information intensive R&D" (AIRD) and related batch/pilot plant level production. Such a culture thrives on an attitude that respects difference and cooperation. It is indeed a synergistic consensus. This culture works as a medium for the institution of network of R&D and production, by way of providing the network with a normative frame, set of (technological) conducts and mission-strategy. A network, we argue, is the necessary institutional and organizational mode for the AIRD and related production.

The significance of AIRD is that with it the R&D and production work together non-sequentially. Ordinarily R&D and production are sequenced and therefore are mostly separately institutionalized. However, AIRD has to happen together with production. This is dynamic R&D-production frontier. Hence, the two together constitute one institution. We argue further that the institution of corporate enterprise is based upon a sequenced and separated R&D, production, trade, etc. With the disappearance of sequence, institutions of big corporate give way to the institution of network. AIRD necessarily requires the institutions of network.

Contemporary legal-banking and other institutions are designed primarily for the operations of a corporate. Notwithstanding this, networks have been in existence (although often languishing) in India and elsewhere. Such network mode encompasses virtually the entire spectrum of technology. These networks could still exist, even without supportive institutional frameworks because culture provided a conducive medium for an array of conducts and norms, for cooperation and for an impressively economic formulation of mission-strategy. The advanced industrial culture provides for these norms, conducts, cooperation and mission-strategies through an attitude of working with AIRD.

Germinal of the institution of AIRD-networks can be perceived to be in operation. Japan especially has brought into a large number of reforms. These are mostly statutes or related changes in the working of institutional finance. These strategies are geared for promoting AIRD and related production in a network mode. Such a strategy is dependent exclusively on the knowledge and skill of the population. Given this precedent, and considering that we lack the institution of network, we ought now to legislate, bring out regulations, promote associations, chart out conducts and norms, and arrive at a consensus, especially as regards what our mission-strategies are, such that we too can forge network institutions and promote AIRD and related production. That is how we may capitalise on what we have already initiated.

Consensus and Strategy

There are a number of doubts however. Do we all share this belief in the necessity of mastering this culture ? Even if many of us do share such a belief, do we then agree on the connotations of this belief? That is, do we indeed agree on not only the meaning of this culture, but also on the ways and means of achieving it? Perhaps we do not share a consensus.

Let us begin from this word "consensus". A textbook definition of consensus (as is ordinarily provided, say, in a book on decision theory) has precise decision-contents and an associated condition that stipulates the foregoing of any other decision-contents by "all" the agents which goes against the contents of the consensus decision. This consensus is therefore not synergistic. It is based on a common minimum core. None of the agents gains in terms of a rise in decision-contents, but the agent gains in terms of an increase in obligation. Consensus in this case increases the obligation and, therefore, acts as a hedge against uncertainties or risk.

There is, however, another view on consensus as, for example, expounded by Maruyama in 1984 and by Gandhi in several places. This view emphasises synergy or the gain in contents. No agent, according to this view, is complete in itself. A consensus does not demand the shedding of non-common decision-contents, but in contrast, allows each agent to increase upon the pre-consensus decision-contents through the addition of the contents that is not available through consensus. Hence this is synergistic. Moreover, the obligation is not on oneself to adhere to the consensus, because through this one gains. Obligation is not meant primarily for enforcing upon the other. This view of consensus too is a hedge against risk and uncertainties, but on a different plane.

This latter view of consensus is also a cultural view. We recognize that we often work through this view. Therefore, although many protagonists of globalization or of advanced R&D are not speaking about a culture of advanced industrial production, we see no contradiction and surely there is a scope for the consensus of the latter kind. In fact, part of our argument will be to show the scope of a consensus between the demands of an advanced industrial culture and the demands of globalization or of advanced R&D, etc. Such a consensus can only be arrived at in a medium of culture. It is in this case an advanced industrial culture.

Consensus provides a framework to the mission-strategy. There are levels of strategies. The obvious strategy, we may note, is at work at a local level such as at Bangalore. In fact, in one instance, a committee of one-hundred members, drawn from multiple agencies, worked together on an airframe. Doing research and designing, fabricating and testing, etc. — all the steps involved consensus. However, the consensus was to be of limited duration and was to be over with the achievement of the mission goal. This is an example of the immediate level of mission strategy.

At a higher level the consensus needs to be achieved on the framework of institutions and on policies. To take an example, we may note the difference in strategies adopted by the American and Japanese governments. The American way of enabling the innovation on radically new advanced technologies had been through defence funding. The defence funding, and not the competitive big corporations, has almost always provided a fillip to the R&D and its productionization. The shaping of larger economic interests, diffusion of innovation and cultural adjustments to the new ways of working took place as a follower. In Japan, in contrast, cultural alignments, economic restructuring, diffusion of innovation and shaping of economic interest played the role of the leader in almost all radical technological innovations. Defence strategy played a role of follower. This policy of "from economics to defence", has enabled Japan now to go for aerospace-commercial first and aerospace-defence gets now, therefore, a social and economic platform from where it can be launched whenever there is a need for that. This latter strategy cannot be worked out through governmental funding but has to be necessarily worked out through consensus, etc. We have observed that the aerospace in India has opted for this latter route of "from economics to defence". India is, therefore, going in for a consensus-driven mission strategy.

The third level of strategy formulation is at the intermediate or organizational level. The corporate protagonist's talk on strategy refers to this. It indicates by strategy a goal. Setting a goal is making a strategy in this view. Goal then selects problems, whose solutions are dependent on the performance. Therefore, this view talks of strategy as a goal-setting, and performance as a problem-solving. In contrast, if the problem merges with the goal, what we get is an objective or specific good for which strategy is to be made. Therefore, in this second view, a strategy is for a limited purpose and is to remain operational for a short-period. This strategy is, therefore, contingent and we have named it mission strategy. We argue for a mission strategy. We observed that contingent consensus was arrived at amongst the concerned agents, and a temporary time-bound mission strategy was adopted in certain activities in the aerospace area. We have seen how consensus may work for mission strategy at three levels of varying durations. First, and perhaps for the shortest period at the immediate local level ; second, for intermediate duration at the enterprise or inter-enterprise level, and third, for even longer duration and often as a policy at the country or global level. In case consensus is required for unlimited duration, strategy is then substituted by conduct and norms.

Advanced Engineering and Information Intensive R&D

Our second task is to lay out the connotations of this industrial culture in terms of "advanced-engineering and information intensive R&D" (AIRD). Surely we shall have to describe the distinctiveness of AIRD, if any, from the commonly understood term R&D. Such a distinction, we shall argue, is present at all the levels of culture, content of R&D, the relation between R&D and production, and at the level of the institutions. Hence, the undertaking of AIRD ensures the working of an advanced industrial culture.

This distinctiveness of AIRD, we argue, is that in this the R&D cannot take place in the absence of production. Hence, R&D ceases to be separate and independent of production. Production too takes on a new appearance. As a result, sciences or the theorization of science too cease to be prior to R&D, or especially to the engineering and designing. Once there is no primacy of science-theory to the engineering-designing, and consequently no primacy of R&D to production, we arrive at the working of a culture where knowing things and doing things are but the same. Hence, culturally, hierarchies based on the hitherto existing notions and institutions of primacy, break down and give way to a cultural cooperation. This culture of cooperation is reflected in not only the contents of science and engineering disciplines, but are also reflected in the new institutionalization of training and designing.

Institutions of AIRD are based upon a new consensus between R&D and production. We argue that this is a network-institution. Distinctiveness of this network-institution from a corporate is significant. A corporate is a co-ordinated routine activity, undertaken by rational non-owner managers, as propounded by Chandler in 1977; by Horn & Kocka in 1977 and by Nelson & Winter in 1982.

A corporate is also a hierarchy that has substituted the market for itself. Why should not the AIRD be substituted by a hierarchy ! We argue that AIRD is based on a synergy of consensus, and cannot be substituted by a "merger" achieved through hierarchic coordination. This institution of AIRD cannot be defined as either a corporate or a market, which the corporate seeks to substitute. This is the institution of network, which is a non-core non-routine consensus, that runs through a specific type of strategy, structure, and conduct.

Our next argument is that while there exist a set of institutions for the working of corporate and companies, there do not exist as yet much part of the institutions necessary for the running of a network. The network, as argued above, is based upon the consensus between AIRD and producers. This being a new mode of cooperative production-R&D, cannot be rendered the full scope of growth and operation under the institutions of company law, laws on monopoly, laws and instruments of banking and of capital market. These latter institutions are nearly a century old, and are needed to be substituted as the studies by Horn & Kocka, 1979 ; and by Pohl, 1988 show.

Finally, summing up the above three arguments, namely, (a) AIRD is dependent on production, neither as the precedent nor as the consequent, but as simultaneous; (b) this simultaneous working of AIRD and production takes place through network and cannot be encompassed under either a corporation or the market; and (c) this network is in need of supportive institutional structures; we may observe that these supportive institutional structures are but tacit approval of an advanced industrial culture. This culture, similar to any other culture, is reflected through consensus, structure and conduct. Advanced industrial culture, we argue, achieves this consensus through the path of strategy and, therefore, mentioned above the working of this culture takes place through strategy, structure and conduct. In the following sections we elaborate upon this theme.

Simultaneity of AIRD and Production

The advanced engineering and information intensive R&D (AIRD) is such a process as would require first an attitude to work, second a concurrent process input and third a platform to implement its results. This attitude is formed by way of imparting a rigorous drilling in advanced engineering practices and in practice that employs information. Information is no longer an input, as is the case otherwise, if the information is culled out, for example, from a patent search. A standard example of this attitude should be the way an experienced CNC machinist-programmer copes with a job. In fact, we are emphasizing the (ethnology) of advanced working in order to understand this attitude. It is also important to recognize that such an attitude constitutes the first element of the culture we are talking about.

Beyond this attitude, in the working out of the process of AIRD, information from the AIRD process itself continues to remain as an input to advanced engineering of AIRD. This is an iteration, where AIRD information constitutes the process input to the AIRD engineering work. For example, a wind tunnel test design for an aircraft requires AIRD information to be followed up by a wind-tunnel testing which requires AIRD engineering. Lastly, an advanced engineering platform is needed to implement the AIRD result. However, this is not indeed the end of AIRD. A platform is actually a pause, from where a batch is produced.

The batch-production or pilot-plant production, once again, forms the input to the attitude of AIRD, which again would produce another new batch. For example, take the case of smart structure engineering, micro engineering, or intelligent mould making. A micro-engineering would require an attitude to begin with, and an advanced engineering and information to carry on the AIRD, and finally a platform such as an aircraft or a robot for the implementing action. This platform too is the subject matter of another AIRD undertaking.

AIRD is distinctively different from other earlier modes of R&D or of production. Ordinarily understood, R&D is separated from production. There are two standard streams of R&D, first incremental R&D undertaken often at the shopfloor or as in-house, and second the non-incremental inventive type, undertaken often at the universities, the publicly funded research organizations, etc. In either case there is a time-sequence of operation. An incremental R&D would often succeed the production proper. For example, Chandler observed that corporate bodies undertook R&D in the form of incremental R&D following, almost always, a stabilized production. In the inventive non-incremental R&D, the R&D precedes the production and has often to be interfaced with production through design engineering, pilot plant, prototype development, pre-production testing, etc.

The significance of sequencing is that the model runs under the assumption of a separation between R&D and production. In fact, we may refer to two significant examples. The first is the evolution of machine tools, from out of a separation between machining and machine-making activities. The recent developments in varieties of machine centres is an outcome on this earlier development of machine tools. We are now perhaps not very sure whether this separation was indeed necessary or whether it was more of an outcome, of a specific social process as shown by Noble in 1986. Indeed if we consider the recent variants of innovations, from the FMS to group technology, or the cells, etc. and also notice that each of these machine tools caters to specific types of production organizations (e.g., FMS for big corporate, cell for the SME), then surely we may infer that such a separation was largely institution-driven. Similarly, we may take up the example of modularization, especially of the process itself in a large processing plant (for example, chip manufacturing). Initially such a separate module of the process itself could be worked upon in order to achieve incremental inventions. Incremental changes in such a process module, say in chip production, could then be put onto the sequence of production either preceding or succeeding the manufacturing proper as was suggested by Zysman in another situation.

Looking at such a sequence, we notice two facts: (1) making of a sequence cannot be justified on the ground of engineering, science or perhaps even of economics; a sequence came up owing to the institutionalized interests in the history of certain set of countries; and (2) a sequence allowed two types of seriality, first a temporal seriality and second, an organizational/divisional seriality. Such serial unit operations allowed for the emergence of big corporate through vertical/horizontal integrations. It is rather important to note that the economy in transaction cost achieved through integrations on a sequence, requires necessarily the prior presence of a separated sequence in production, R&D, sales, etc. as was suggested by Williamson. In the absence of such a sequence, therefore, there is all the likelihood of the non-emergence of a corporation.

In AIRD, we observe an absence of such a sequence. We observe, as for example in our field studies at Bangalore on the aerospace segment, that there is an absence of a sequence between the R&D of AIRD and manufacturing. This lack of separation between AIRD and production, is reflected firstly through the temporal simultaneity of R&D and manufacturing and, secondly, through the role-concurrency of AIRD and manufacturing. Temporal simultaneity requires that AIRD can neither precede nor succeed manufacturing. In this the advanced engineering and information intensive R&D takes place simultaneously with the production. As a result, we can no longer consider the R&D as an "input" to the system of production. Moreover the role of AIRD alone is no longer contemplative. Production too is contemplative, futuristic, prognostic. Overall we observe the emergence of a situation where R&D and production are no longer ordered in a sequence (on time and on role), but do appear together.

Cultural Connotations of Simultaneity

AIRD presupposed the presence of an attitude. Such an attitude is formed through a long association with advanced engineering practice and with the use of structured information. It is therefore, a socialization. At our field, at Bangalore, we observed an attitude at work. Perhaps the initiating group shared an attitude, that to begin with was formed by the discipline of aerospace engineering. However with time, this attitude overstepped the boundaries of the discipline of aerospace engineering. It was now to be inculcated through such multifarious practices as information intensive designing, AIRD intensive material for using, designing special machines, information intensive designing of advanced engineering experiments, micro engineering, parallel computing, designing of standards, etc. In fact, this has involved in and around Bangalore, such varieties of people as are vendors from SME, technicians and mechanics, engineers from a large number of disciplines, scientists and academicians, *et al.* As a result, this group of people can now share the rudiments of a common technical language, can perceive and conceptualize both a problem and its solution in a way that is compatible with advanced engineering and intensive use of structured information, and finally share a culture of work. This all, together, forms the attitude.

In a sequence, where there is a separation between R&D and production, there is a culture of "performance". We may recall that the protagonists of a corporate (of a managerial hierarchy, etc.) claim that a corporate is a rational and economic organization on the market. This corporate has to have a sequence order between R&D and production, divisional structures, corporate strategy and most importantly a drive for performance. In the culture of the corporate, performance occupies the central place. Performance signifies, the "efficiency" of an operation without any regard for the goal of the operation. It is therefore, a technical efficiency. Similar to any other notion of technical efficiency, the "performance" does not have any consideration for the goal. Goal is the concern of strategy. Therefore, a culture of performance, which is the culture of the corporate, separates out the concern for the goal from out of the operations of a participant. This participant can act only for an efficiency.

In contrast, in AIRD and related network of production centres (factories, for example, or a batch processing pilot plant), there is an absence of sequence and there is a milieu of an attitude. The absence of sequence or of division necessarily demands from each participant that the participant has the goal or a strategy. Therefore, the participant cannot sit back with performance alone. Now, as argued above, the goal in a network is a cooperative dynamic-goal. Such a goal can be arrived at through consensus, cooperation, norm, and above all through an attitude. The culture of a network is thus to have an attitude such that one can cooperate, can arrive at a consensus, can frame and abide by norm, etc. We may notice that if the concerns of performance are to be quicker, efficient and economic then the same objectives are satisfied, in an equally satisfactory manner, by the attitude prevailing in a network. This attitude enforces the participant to abide by the norms, to cooperate and to enforce on oneself the dynamic-goals as set by the consensus. The culture of AIRD and related network of production centres is of an attitude, which is sharply different from a culture of the performance prevalent in the corporate.

Attitude and the absence of sequence have rather significant implications for the training, etc. Firstly, the standard dichotomy between theory and practice-training becomes redundant now. Since AIRD itself has given up its separation from the manufacturing, the training in a network should necessarily be a synthesis between theory and practice-training. This involves a new design of training and a new content as well. Secondly, the content of a training is now indeed of an interdisciplinary nature. In particular the separation between abstract theory and practice breaks down resulting in the disappearance of the primacy of abstract theory. With the loss of the primacy of the abstract theory, the theoretical disciplines too begin to loose the primacy. The theoretical disciplines begin to incorporate more and more practical experience, and often new disciplines emerge. Such disciplines are extensively dependent an experiments, and borrow heavily from other disciplines.

Thirdly, engineering tends to become more of an art as was suggested by Rosenbrock in 1990. Since changes in disciplines, as argued above, are brought forth primarily by experiments; this devising of experiments appears as more and more of an art. In fact, quite a few new methods of training in European countries recognize the importance of art as was indicated by Cooley in 1990, and in Europe they are bringing training closer to the designing phase of the product or process as was pointed out by d'Irribarne in 1990. The institution of AIRD (which is a non-sequential network) is doing away with separate and abstract R&D, and is closely approximating the experiment — where a synthesis between art and systematic observation is taking place. Training in AIRD mode is a training in designing, experimenting, systematizing. Therefore, such a training is not organizationally separable from the R&D, batch or pilot production, or the devising of new machines, etc.

Network in Lieu of Corporate

Culture provides the milieu indeed. A milieu, however, does not act. Active roles are undertaken by organizations, and the institutions. It is at this organizational and at the institutional levels that we observe the most striking distinctiveness of a network at play.

Recalling what we have argued above, we may summarize by saying that a network is not unique to the affair of AIRD alone. The network mode that is at work at Tirupur in South India is an interesting example. Tirupur production units, individually each is small and the units in general do not employ sophisticated capital machines, although use of latest information tools, CAD, communications mode, modern infrastructure including warehousing are prevalent. Therefore, the preconditions of a hierarchic corporation are materially present. However, what we observe at work at Tirupur is not a corporate but a network. Similarly networks of master traders-craftsmen are seen to be operative in such technologically backward areas of operations as textiles, silk, etc. Earlier for example studies indicate the network at operation amongst the Marwari community, The very powerful presence of Chinese network (the so-called diaspora, that of Hui community) in South-China and South-East Asia is surely worthy of attention. In fact, Zaibtsu's role in the formation of the later Japanese conglomerates (which are not surely a twin of the American corporate ; for example Mitsui cannot be equated with Unilever or Dupont) cannot be underestimated. In fact, even in the heartland of corporations, namely in Europe, we notice the emergence of network as has been pointed out Johansson, Karlson & Westin in 1994. Piore in 1994 categorically drew our attention to the contemporary American evidence whereby one must look sceptically at Chandler-Williamson doctrine of the permanence and universality of big corporations. Therefore, it is time indeed that we recognize the distinctive institution of a network. We do not wish to explicate in detail the distinctive features of a network. However, certain features may be pointed out. Firstly, we notice that in a network trade is not separated from production. Secondly, a network does not operate under a "routine" of either the administration or the production, and this network is often non-core. Thirdly, a network operates for a time-bound (or contingent) strategy, which is in the nature of a mission. Fourthly, a network is maintained through conducts (or more properly technological conducts) of each of its participants to any other in the network. In short, in a network we experience cooperation as the central theme.

These being some important features of a network, we may notice that while such networks are present sometimes only in many spheres or regions (such as at Tirupur), they appear to be necessarily present in a situation of AIRD and related production, such as in aerospace or at Bangalore. We are not going to offer any explanation as to why in many industrial sectors networks are not significant or are not strongly present. Suffice it to say for the present, that contemporary institutional arrangements are heavily in favour of an enterprise-corporate structure (enterprises of Sombert-Weber type, corporate of Chandler-Williamson type) and these do often put up severe odds for a network to flourish. We may, however, note that in AIRD and related production, network appears to be the necessary organizational and institutional framework.

Recalling AIRD features where there is a simultaneity between R&D and production, it is easily understood that there cannot be a divisional structure. Hence we cannot have a multidivisional firm here. Again AIRD engages itself with such a R&D-production, that each production set is as it were, an experimentally (driven and) devised batch. In the next, production set will have improved or changed itself so much as beyond the scope of an incremental change. Therefore, in this dynamics of production the system cannot retain a core. A core requires stabilized routine production. In AIRD-production there appears, therefore, a network in the place of a core.

This dynamics of production can be maintained only if there is a mission. Since, there is an absence of a core, production-set can be maintained only if there is a mission. This mission drives the dynamics of production-sets. We have been using, for convenience sake, the term strategy for the meaning of a mission. Strategy, although is used by the protagonists of corporate in a rather different and somewhat contradictory sense. We may use the term mission strategy in order to convey our intended meaning. The network will, therefore, require a mission strategy. A mission-strategy is always contingent, that is a mission strategy is always for a limited number of years and for a limited, specific purpose (or good). A strategy, such as that for the profit, and as used by the corporatist, is unlimited in time and inspecific in purpose. Profit as a goal cannot tell how best to attain it and for how long one should work for that. Strategy for profit must therefore have criteria on performance. However, the mission strategy is entirely positive (without a blemish of natural law). It is entered into, for a specific good, for example to attain better than world class productivity in leather tanning ; and for a specific time, for example, next seven year, to achieve that.

A mission strategy will surely require of the members of the network to arrive at a consensus. In a corporation, a making of strategy is the onus of the top-management (or the Chandlerian entrepreneur). A hierarchy alone, in an authoritative mode, can implement such a strategy. It is also known to us how many experiments with quasi-authoritative, quasi-hierarchy attempted to grapple with strategy formulations and performance implementations in the republics of Eastern Europe, especially in Yugoslavia. Attempts at democratization of the strategy formulation have thus failed. Therefore, for the societies of Western Europe and for the USA, corporations became the only centre of authority. A democracy in the society had to be complemented with the divided sovereign — the hierarchy inside big corporations. In contrast, mission strategy requires consensus. A consensus, as argued at the beginning of the study, is synergistic, only by way of fulfilling certain conditionals. These are, for example, constructive or additive cooperation, obligation on oneself rather than on the other party, etc. A consensus, such as on network mission strategy, runs on this kind of obligation. Any violation of the consensus alone can invite the extra-network authority, for example, a regulator or a government department, to adjudicate.

A consensus requires an initial point. We have seen how the attitude acts as the medium or the milieu. In this medium of attitude, consensual-process on a mission-strategy is initiated by the conducts. In fact, what we observe in AIRD, for example, at our field, is the technological conduct. A conduct is positive and is not derived from reciprocal conduct of others. It is based upon both the systematization of a very large number of behaviours, and on the injunctions on behaviours. Thus, the contemporary conducts are but the systems of mutual behaviours. We may wonder how in a dynamic situation, such as with the AIRD and related production, a system of conducts can cope with the ever emerging demands on the behaviours. Field observation and our reasoning dictate that, injunctions do play a crucial role here. Injunctions are like bind on channels, such that new demands on behaviours are disallowed certain modes, leaving free a vast field behind for the behaviours to take upon shapes. Moreover, systematized information on conducts do also appear as "types" or as conducts. In short, conducts are the living force of networks. Even an absence of mission-strategy would not allow the network to collapse — conducts run the affairs of a network even while mission strategy is absent.

Networks, therefore, can stand in lieu of the corporate. Evidence suggests the operations of network, perhaps, at all the levels of the spectrum of technology. However in AIRD and related production, the network appears to be the necessary organizational and institutional mode. Moreover, a network also appears to be culturally favourable to the business management in the East as was argued by Hamashita in 1992 and by Negandhi in 1985. Indian culture and Indian social organization seem to be especially favourable to the running of a network.

The Missing Framework — Absence of Institution

The thriving Indian business networks, for example of the banias, suggest that despite the absence of certain institutional frameworks, the networks could successfully retain the business. There must have been therefore, some other non-official institutional frames at work. We may therefore, accept two institutions ; first, an official and mostly formally structured institution and, second a social non-official non-formal institution. An example of the former is company law, banking law, etc. An example of the latter is cultural and social norms, such as values, trust, cooperation and community, etc. Our first observation is that while the former in general has strongly disfavoured networks, the latter has but favoured network.

Horn in his study on the history of business laws in Germany, Great Britain, France and the USA concludes that company law contributed immensely to the development of organizational structure of modern enterprise. By early twentieth century, the joint-stock company became the most important legal form of organization for large-scale enterprises. Horn summarizes four main aspects : (1) institutionalization of enterprise as legally autonomous unit, (2) division of the decision-making authority from its control within the enterprise, (3) two-fold and changing role of shareholders as entrepreneurs and investors, and (4) some general relationship between company law and the economic growth. These changes allowed easier accumulation of capital from many small investors. However, these same company laws did not provide any regulations for the internal structure (or staff), where general rules of agency continued to remain. Slowly, and somewhat concurrently other legal-institutional structures, especially that of the banking came into place in order more to enforce the incorporated structure of the corporates.

It is doubtful surely if in the absence of these above institutions, the corporate would have emerged at all by dint of the much-proclaimed economic rationale alone. In fact, in countries other than the cultural sphere of corporation, such as in Japan, these legal structures were somewhat enforced during the period of political exigencies. Perhaps, as a result of this we observe in Japan a certain compromise between the corporate and the network as has been shown by Kobayashi and Morikawa, 1986; Mitsuhiro, 1993. In fact, the big business in Japan had an earlier practice of running the business through appointed managers. So did we have a practice of running the business through managers. However, these managers, over time, also turned out to be part-owners. In Japan and in East and South East Asia, varieties of quasi-official bodies came up, more perhaps to make up for the adjustment and also to make good the losses inflicted by the imported framework of official institutions, of the company law — of the banking law, etc.

Examples of such bodies are multifarious. Associations played a crucial role not only in the Japanese prefectures, but also at the principal locales of the hierarchic conglomerates. Internal working of the conglomerates made little use of agency laws. In India, however, in the absence of strong agency regulations, the internal organization of the big business and the factories were institutionalized through general governmental frameworks — for example tripartite industry-wide wage fixation (sometimes even the fixation of such internalities, as the promotion, etc.). The weakness of corporate in these Asian countries has become too well known a fact. In Japan, while governmental agencies such as the MITI or the associations of trades, etc. attempted to retain the benefits of networks, in India something contrary took place. Weakness in Indian system, owing to the derecognition of networks, was attempted to be alleviated not through a working of the system together with the network but through the working of a government-level framework. Indian networks, some of them at least, have thrived notwithstanding this weakness.

Therefore, if our legislators and statesmen would like to pay heed to the writing on the wall, what they need to provide the Indian business with, is the appropriate institutional-legal framework for the network. Such an institution, we argue, is an imperative if we are serious indeed regarding our occupation with AIRD. In fact, there are indicative changes, already brought into the field, by umpteen number of minor amendments to the existing framework.

Japanese experiment with short-term positive laws is an important achievement. The Japanese government have been bringing out, at least over the last two decades, a number of statutes — that are to expire ordinarily after a stipulated period. They brought out two sets of such laws for the development of the information-technology. These were succeeded by similar short-term, contingent, sharply targeted legal instruments — such as the Technopolis law. In fact, this law ushered in the later institutions of science and technology parks (STP). Latest in this series is the 1995 law on science and technology. These laws are providing the prefectures, councils, enterprises, R&D institutions, universities, etc. a corresponding institutional framework. These are providing them with: (1) mission strategies on a contingent basis; (2) a norm-institution whose violation does not attract penalty but perhaps only an outcasting; and (3) an extensively worked out structure of conduct. Therefore these are providing the Japanese with the structures of network-institutions. Moreover, the structures of government, such as the departments, are playing only supportive and secondary roles in this regard.

There are some other significant developments. The experiments with STP, through varieties of organizational structures and institutional arrangements — in order to promote innovation, primarily in the AIRD areas, is a case in point. There are notable changes also in the areas of operations of institutional funds as has been argued by Aoki. All this is but part of the changes, that are somewhat in favour of networks, especially of AIRD and related production. In fact, as a result of these changes, an industrial restructuring of Japan has been going on for the last one decade. The full-set technological structure of the industry is a thing of past now in Japan, which is now intensively working for AIRD and related production, primarily through a reemphasize on skill, education and AIRD network.

Japanese experiments in this restructuring are worth studying. In our field at Bangalore, in the areas of aerospace industries, the AIRD and related production is slowly taking up the network mode. However, there are no institutional arrangements, at Bangalore, comparable to what the Japanese have begun providing. Equally network at Tirupur or in the rural areas of Bengal, are thriving or languishing as is the case, without any institutional arrangements whatsoever. Our present need is to legislate and design proper institutional structures for networks not only for AIRD and related production but also for the backward areas of technological activities.

Medium of Culture—Strength of Institution

This above refers to the rather formal institutional structure. Such structures operate in a medium. In fact, the way injunctions (refer above) are accepted, the way norms and conducts are settled and even therefore the way mission strategy reflects itself in the actions of the network members — all this is the way of the culture. Culture, therefore, is not the norm, nor it is the conduct. Culture provides the values one attaches to norms, conducts.

Culture is the attitude, we have referred to above. Therefore, in as much as an attitude is nurtured and inculcated, so much is the progress of this culture. Culture is not only what we receive as a patrimony, but it is also what we continuously generate, nurture and inculcate. In order that one appreciates AIRD and related production, one must have worked with structured information especially in systematizing and applying the information. Equally one must have worked with advanced engineering. Through AIRD, the participant acquires a symbolic and technical language of comprehension and communication. This indeed is a point of view or an attitude. In the absence of such an attitude or culture, AIRD cannot simply remain operational. AIRD, through its working, generates this attitude — a sort of platform, shared by the participants.

Culture is on multiple platforms. We have seen above how, in our culture, a consensus encompasses the uncommon together with what is common to all. In this synergistic consensus, the platform is thus enlarged. Moreover, the obligation that one would adhere by the consensus is on oneself, rather than on the other party. The direct relevance of such a consensus can be gauged by looking at the great economy, especially in information, that the group achieves, say for example in cooperative production (as at Tirupur), in franchising, in arbitration, etc.

The advanced industrial culture is an attitude to work with AIRD, is a consent to the encompassing of the common life by the advanced industrialization, etc. Such a culture poses some special questions in our case. In a culture of non-synergistic consensus, the culture forces the differences to its death. The common minimum core then becomes the great homogenising force. In synergistic-consensus, however, the synergy is strengthened because there have been and there are differences. In advanced industrial culture of the sort we have been arguing for, differences are respected. A culture of the network is a culture of the difference. Cooperation in network can thrive only when differences exist not antagonistically but with a sense of dignity to one another. In AIRD, we have seen dignity also being attached to the vendor having only simple jigs. Novel and ingenious solutions often come up, from such a vendor, provided that he too shares the attitude of AIRD. Therefore advanced industrial culture in our context does not imply the necessary employment of sophisticated capital-intensive machinery. In Bangalore we have seen attitude at work in the development of parallel computing with but little capital employment. An advanced industrial culture for India is a culture of difference with an attitude of the AIRD.

Concluding Remarks and Further Research

In this article we have defined advanced industrial culture as an attitude of working with "advanced engineering and information intensive R&D" (AIRD) and related batch/pilot plant level production. Such a culture thrives on a culture that respects difference and cooperation. It is indeed a synergistic consensus. This culture works as a medium for the institution of network of R&D and production, by way of providing the network with a normative frame, set of (technological) conducts and mission-strategy. A network, we argue, is the necessary institutional and organizational mode for the AIRD and related production.

Significance of AIRD is that, with it the R&D and production work together non sequentially. Ordinarily R&D and production are sequenced and therefore are mostly separately institutionalised. However, AIRD has to happen together with production. This is dynamic R&D-production frontier. Hence, the two together constitute one institution. We argue further that the institution of corporate enterprise is based upon a sequenced and separated R&D, production, trade, etc. With the disappearance of sequence, institutions of big corporate give way to the institution of network. AIRD necessarily requires the institutions of network.

Contemporary legal-banking and other institutions are designed primarily for the operations of a corporate. Notwithstanding this, networks have been in existence (although often languishing) in India and elsewhere. Such network mode encompass virtually the entire spectrum of technology. These networks could still exist, even without supportive institutional frameworks because culture provided a conducive medium for an array of conducts and norms, for cooperation and for an impressively economic formulation of mission-strategy. The advanced industrial culture provides for these norms, conducts, cooperation and mission strategies through an attitude of working with AIRD.

Germinal of the institution of AIRD-networks can be perceived to be in operation. Japan especially has brought into a large number of reforms. These are mostly statutes or related changes in the working of institutional finance. These strategies are geared to promoting AIRD and related production in a network mode. Such a strategy is dependent exclusively on knowledge and skill of the population. Given this precedent, and considering that we lack the institution of network, we ought now to legislate, bring out regulations, promote associations, chart out conducts and norms, and arrive at a consensus, especially as regards what our mission strategies are, such that we too can forge network institutions and promote AIRD and related production. That is how we may capitalize on what we have already initiated.

This above being our conclusion, we have now to look

forward to formulating some research questions. The focus of our research question is on what constitutes the necessary institution for the promotion of networks, especially in the areas of AIRD and related production. In order to do that we would investigate actual field conditions rather extensively. We should also make comparisons with the developments in East and South East Asia. These together will show us how to promote through varieties of legal-regulatory and other frameworks, the fledgling institutions of network especially in AIRD related areas. Some of the types of questions that seem to be worth probing are,—

(a) role of business associations in disseminating information; effecting norms and conducts and in achieving cooperation and consensus;
(b) intensity of the use of structured information, information interfacing of the enterprise and use of shared information in a locale, especially in designing;
(c) user profile of advanced engineering machinery, advanced engineering practice in an area or for a mission; and the nature of AIRD training;
(d) short-term legal/regulatory measures that have been either demanded or implemented recently; and
(e) Role of institutional finance — nature of and changes in control; nature of recent changes in cross-equity holding and relation of traders or its associations to the changes in manufacturing.

BIBLIOGRAPHY

Chandler, A.D., 1977, *The Visible Hand: The Managerial Revolution in American Business*, Cambridge, Mass: Harvard University Press.

Cooley, M., 1990, *European competitiveness in the 21st century: Integration of Work, Culture and Technology*, FAST, Commission of the European Communities.

d'Iribarne, A., 1990, "Advanced Technological Forms which reflect the Cultural, Economic, Educational, Geographical and Social Reality in Europe" in Cooley, *op. cit.*

Hamashita, T., 1989, "Tribute and Emigration: Japan and the Chinese Administration of Foreign Affairs," *Senri Ethonological Studies, 25.*

Horn, N. & J.Kocka, 1979, *Law and formation of the big enterprises in the 19th and early 20th centuries*, Gottingen: Vandenhoeck & Ruprecht.

Johansson, B., C.Karlsson & L.Westin (ed.), 1994, *Patterns of a network economy*, Berlin: Springer Verlag.

Maruyama, M., 1984, "Alternative Concepts of Management: Insights from Asia and Africa" *Asia Pacfic Journal of Management*, January.

Negandhi, A.R. 1985, "Management in the third world" in Pat Joynt & M. Warner (ed.) *Managing in different cultures* universitetsforlaget As.

Nelson, R. & S. Winter, 1982, *An evolutionary theory of economic change*, Cambridge, Belknap Press.

Noble, D.F., 1986, *Forces of production; A social history of industrial automation*, New York: Oxford University Press.

Piore, M.J., 1994, " Corporate reform in American manufacturing and the challenge to economic theory" in T.J. Allen & M.S. Scott Morton (ed.) *Information technology and the corporation of the 1990's : Research Studies*, New York : Oxford University Press.

Pohl, H (ed.), 1988, *The concentration process in the entrepreneurial economy since the late 19th century*, Stuttgart : Franz Steiner Verlag Wiesbaden GmbH.

Rosenbrock, H., 1990, *Machines with a purpose*, Oxford University Press.

Zysman, J. & Tyson, L. (ed.), 1983, *American industry in international competition*, Ithaca and London: Cornell University Press.

14

Technology Transfer and Human Resource Development

B.R. Virmani

In today's context, technology is the key to the growth and development of any nation. Some countries, because of their innovative practices and inventions, have been able to develop technology at a much faster pace resulting in better living standards for their people. However, there are a vast number of less developed countries which, for various historical reasons, could not keep pace with technological developments. Also, with the changes in the income patterns, communication amongst the nations of the world and people, it becomes a futile exercise if each country every time develops its own technology from scratch. While some countries develop their expertize and technology in certain areas, others have different kinds of skills and resources to offer. Therefore, it becomes imperative that technology is shared for achieving better living standards so that all the countries gain from the same. With the globalization concept, the competitiveness has increased and many organizations lack the resources to develop technology in all aspects on their own.

The need for continuous innovation specially in high technology industries' alliances plays a major role in order to fulfil the technological gaps, increase the interconnected product range, to tap new markets and meet competition, to service the customers better and above all to save time in developing new products and process technologies.

The need for entering into alliances between one or more organizations is to meet basically the emerging challenges and bridge the gaps in some of the aspects like R&D,

manufacturing, marketing and at times for general purposes. The type of alliances entered into would directly reflect the objectives to be achieved.

Types of Alliances

R&D Alliances : Licensing agreements, cross-licensing, technology exchange, visitation and research participation, personnel exchange, joint development and technology acquisition investment.

Manufacturing Alliances: Original equipment manufacture, second sourcing, fabrication agreement and assembly and testing agreements.

Marketing and Service Alliances: Procurement agreements, selling agents, servicing agents.

General Purpose Tie-ups: Standards coordination and joint ventures.

Technology Transfer and Assimilation

Technology transfer is the utilization of an existing technique in an instance, where it has not previously been used.

Absorption and assimilation of technology by the recipient organization/country is a true yardstick of the actual transfer of technology, otherwise it will remain only as a transplant of technology for manufacturing/production purpose without developing the appropriate capabilities of recipients of technology.

Therefore, for technology transfer it is important that the recipient organization is able to assimilate and absorb the technology, which in other words means that the organization has developed capacity to utilize the technology on its own. Technology assimilation should be a major consideration of recipient countries in order to enhance and derive maximum benefits from a technology alliance. The role of multinational corporations as transferers of technology apart from other alliances is a major source of latest technology and know-how. True alliances should reflect equal gains to both parties.

Role of HRD

One of the crucial factors of technology transfer is the

capability and skills of human resources to receive, assimilate and absorb the technology. Also, due to various environmental constraints (social, economic and political), there could be resistance to receive or absorb new technology which may reduce the competitive advantage of the recipient organization/country but is defended on grounds of meeting social objectives. Employees fear restructuring as a potential loss of jobs or additional cultural stresses and thus oppose a technology alliance. Many a time the people within a firm are in a position to make or break an alliance. Therefore, advance preparation of the employees to accept technology alliance is absolutely essential.

This study discusses some of the issues connected with transfer of technology and the preparedness of the organization to assimilate the same including some of the factors which help or hinder the process of assimilation, with special reference to the human resources aspects. The second part of the study examines these issues in the light of case-studies of two organizations in India—one, a large petrochemical company under the public sector and the other, a private sector electronic (information technology) company. The study of these companies was undertaken because of the deliberate efforts being made by them to assimilate the technology with special emphasis on HRD. A few suggestions are also made to increase the capability of developing countries specially the human resource aspects for the success of technological alliance for mutual advantage of the involved parties.

Process of Assimilation

For proper assimilation of technology, some of the important components of technology play a crucial role. These components are object-embodied technology such as facilities for testing, production and manufacturing design equipment; person-embodied technology for providing the database to support design, testing, analysing and monitoring; and institution-embodied technology to integrate the other three components, organization, government policies, etc.

These have been referred to as Technoware, Humanware,

Infoware and Orgaware by the technology Atlas prepared by the United Nations Economic and Social Commission for Asia and the Pacific (UNESCAP). The major steps in technology assimilation could comprise identification of the needs and sources of finance for investment, market demand and supply, technology forecasting, availability of technology sources, choice and negotiation of technology, production, commercialization, adoption and upgradation.

Among other factors, the human component plays an important role in the entire assimilation process with infoware providing the basic support to assess the technology, to make the appropriate choice, to negotiate, etc.

The success of the technological alliance is also to a great extent dependent upon the capability of the recipient organization to receive technology. At times, the stronger partner may not even be keen to fully train and increase the capability of the other partner due to fear of losing its competitive advantage. They may offer training only in operations and not in the process with the result that even for a minor defect or rectification, the experts from the technology giver/transferor need to be brought in. This blocks the entire process of assimilation, and in some cases eventually results in the breakup of the alliance as the recipient partner finds itself in a much weaker position.

The recipient organization does not adequately prepare those who are identified for training with interorganizational roles. Without a proper understanding of what the organization is trying to achieve, the learning period is not fully utilized to bring in maximum knowledge and learning. Learning is also to be monitored and recorded systematically. Normally, the documentation system is poor and while no records are maintained, cross-checking and verification becomes difficult. Assimilation and absorption are highly dependent on recording the actual learning i.e. the process, know-how, etc.

Characteristics of Indian Organizations for Alliances

(i) With the opening-up of the Indian economy and prospects of multinational companies seeking entry, the alliances with Indian companies could offer mutual advantage. Some of the major benefits which Indian companies can offer are,—

(a) large domestic market;
(b) wide distribution network;
(c) reasonable manufacturing base/excess available capacity;
(d) reasonable level of manpower skills and expertise;
(e) relatively cheap labour;
(f) outsourcing of components at lower cost;

(ii) *Increasing Trainability of Human Resources:* With the changes in technology, the skills and knowledge component of the human resources have to be constantly upgraded. Otherwise the assimilation or absorption process gets weakened. The technology Atlas (UNESCAP) defines these stages of sophistication of increasing human abilities starting from operating, setting up, repairing, reproducing, adapting, improving and innovating abilities. All these call for strengthening not only their skill levels at later stages, but from basic and elementary education level increasing their trainability component assumes much greater importance than training persons for a particular skill; the requirement of which may only be temporary; therefore, there has to be a blending of the basic and technical education with on-the-job training at organizational level.

The trainability traits include reasoning skills (development of cognitive ability); development of a scientific attitude; mastery of appropriate language; development of communication skills; development of interpersonal skills and development of self discipline and responsibility.

(iii) *R&D Capability in India:* The involvement of R&D in the process of technology transfer, upgradation and innovation has been limited. Though most organizations do have R&D personnel, yet these are mainly to avail themselves of the taxation benefits offered by the government. R&D in some of technologically advanced countries plays an active role in receiving technology, unpackaging technology and its transfer to operations. However, in India with a few exceptions, R&D has restricted their role.

(vii) *Documentation:* Another factor which at times hinders the process of assimilation is the weak retrieval system due to poor documentation. There is no proper system of documenting and recording information about the various technological changes or modifications brought about in the process from time to time. While negotiating for alliances, at times it is found that the Indian organizations are unable to provide the suitably updated documents of their existing technology capabilities.

(viii) *Functional Coordination :* For total assimilation of technology it is essential to have proper functional coordination among operations, R&D, marketing and human resources development. The coordination between human resources and other functions is important especially with regard to development of appropriate manning norms and identification of training need for handling and absorbing the technology. Similarly, R&D, marketing and operations need to work very closely to understand the process and ensure constant upgradation.

The experience of other countries, especially South Korea, indicates that understanding of the process aspects is much better when R&D professionals are fully and deeply involved in the entire transfer process. In fact, in some organizations in Korea, technology transfer takes place to R&D department and then R&D in turn transfers it to operations. The advantage of involving R&D right from the beginning is that they understand the process aspects and, wherever required, can consider upgrading the technology on their own.

(ix) *Infrastructure:* Suitable government interventions in terms of providing infrastructure services like transportation, communication, utilities, legal framework, financial services, material testing laboratories, setting appropriate standards, etc., are indirect factors affecting the process of technology assimilation. There is now a definite need for industry-government collaboration to improve and develop these infrastructural facilities by providing appropriate incentives to industries.

Apart from training of the personnel for operations of the machines, the technology transfer should include preparation of indigenous personnel to operate and maintain some of the relevant essential infrastructure facilities.

(x) *Human Resource Readjustment:* The government policies with regard to trade union regulations of employer/ employee relations could play a significant role in encouraging foreign investment, transfer of technology and technology assimilation. Any technology change or upgradation could disrupt the labour. Development of micro and macro level policies with regard to handling such labour-related issues such as redundancies, skill changes, and redeployment policies are essential. In this connection, at the micro level each organization has to work out its own readjustment policies, and at the macro level the government has to ensure a conducive legal framework to protect the larger interest of all the parties concerned.

In today's environment in India, the flexibility to redeploy persons due to technological changes is limited because of stiff resistance from unions and associations. This to a large extent affects the assimilation and absorption of technology. Any alliance may create fear of restructuring resulting in potential loss of jobs or even additional culture stresses. One of the methods of reducing this problem would be adopting a participatory approach with regard to technological changes and working out a rehabilitation plan for each employee affected by this change. In Germany, their codetermination law stipulates that the employees have to be informed in advance about any major technological changes (being envisaged) and then the management has to go through a process of the rehabilitation plans along with the employees, representatives to work out a detailed joint rehabilitation plan. It is believed that this process has reduced the resistance to technological change in Germany as compared to some other countries like Great Britain or India.

(xi) *Lack of Technology Assimilation Policy:* In many cases, technology assimilation has not been defined by the companies as one of their missions or objectives. The result is that the organizations concentrate mainly on technology transplant and meeting production targets rather than on developing a conscious long-term strategy of technology assimilation.

(xii) *Equity Participation:* Indian partners in most alliances have found that, due to their poor technology assimilation process, they have had to compromise on their share in the total equity participation. The stronger partner (usually a foreign partner) gradually imposes preconditions for bringing in the latest technology into the venture by increasing its stake in the alliance where it can emerge as a majority shareholder.

(xiii) *Understanding Cultural Diversities in Alliances:* While transferring technology, there is always a tendency to transfer some of the managerial practices and systems also, specially from the Western countries to India with the belief that these management concepts are universal and can be grafted to on the Indian situation. However, whenever such grafting is enforced, it is found that such enforced management practices do not work. In such situations, the organizations are forced to adopt an "Indian" approach by arriving at various compromises and adaptive changes, at times resulting in frustration (Virmani and Guptan, 1992). This also produces a distinct hiatus between the professed policies and the actual practices followed in the Indian environment.

Theorizing that is suited to the essentially Western industrial infrastructure, and foreign or western management

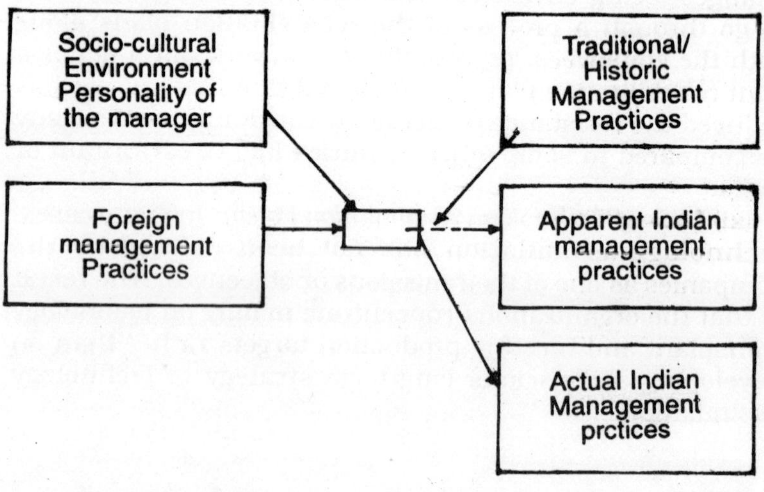

practices would work in Indian conditions too, provided policies are evolved and adequate plans made. The enforcement of such policies results in a strange dualism in Indian management practices, one, which is professed on paper to be western and the other which is actually followed. The following presentation expresses the situation in the Indian management (Virmani and Guptan, 1992) :

While it may be easier to transfer and assimilate the technology know-how, the transfer of managerial technology is a more difficult process because of the cultural gaps involed. For the successful functioning of the technology transfer it is crucial that proper adjustments are made in the managerial technology to suit the Indian environment and culture. Some of the human resource-related policies, the manning norms, the production yardsticks, the promotion policies, the employee appraisal systems or even the salary structure must be adjusted to suit the Indian environment.

In Indian organizations there may be certain limitations on human resources policies, for e.g., there is a pressure to respect seniority or give preferences to certain categories of employment (e.g. reservations). In such a situation the employee appraisal system may be used mainly for developmental rather than promotional purposes. Another crucial point here in regard to growth is the Indian expectation based on the strong belief in seniority. On the face of it, it appears to rub against the grain of the purists who believe that merit alone should prevail. The perpetuation of the conflict arises from the lack of any planned growth or development and going against the character of Indian organizational expectations.

In such a situation the organization has to ensure that the seniormost is also made meritorious to the extent possible through appropriate organizational training and developmental plans. Similarly, the hire-and-fire policies will not be workable or acceptable in an Indian organization. Systematic career planning right from the entry of young people into the organization would, coupled with planned training and development, ensure a satisfying growth for both the people and the organization

Case of Process Industry Limited

Process industries Ltd. (PIL)

PIL is one of the large chemical process industries in India with a sales turnover of Rs. 2279 crores (1 crore =10 million). In 1989, PIL was rated as one of the best performing chemical companies in the world and in 1991 was placed in the top ten.

The major PIL products comprise petrochemicals (plastics, synthetic rubber, synthetic fibre, industrial chemicals), advanced engineering plastics, catalysts, absorbents, wire and cable compounds, corrosion, consultancy and external training contracts. It has around 30 plants spread over three locations. Most of these plants were set up with technology transfer agreements with some of the largest technical firms in the world. It has almost 30 technology transfer agreements—mainly from the USA and Europe. In addition, PIL has joint venture agreements with Pasteur Merieux of France for human viral vaccines; with GE Plastics BV Netherlands for advanced engineering plastics. They have entered into joint ventures with large Indian companies for port facility. PIL is also in the process of setting up a joint venture in Saudi Arabia. It has an agreement to train technical personnel from Iran.

At the initial stages of establishment, the PIL as a policy brought the technology of lumpsum payment basis without any equity participation with a view to subsequently developing the technology themselves. In most cases, the technology was brought as a packaged deal and no reverse engineering was undertaken, and the agreements had restrictive clauses with regard to technology upgradation. The period of agreements were long, ranging from 6 to 15 years.

Before 1991, the investment through equity participation of the foreign partner was discouraged except in very specialized sectors. However, after 1991, the situation had changed and PIL also entered into alliances based on 50 per cent equity participation. Recently the company has gone in for equity participation with appropriate transfer technology alliances with GE Plastics and Pasteur Merieux.

The development of technical capability within PIL and also its past performance were able to attract foreign investment which in turn helped PIL to obtain the latest technology from its counterparts.

Nature of the Latest Joint Venture in PIL

However, in the more recent alliance, PIL is attempting to bring in a totally new work culture and has tried to rectify some of the problems discussed earlier.

This new alliance was entered into on the premise that the very latest technology in advanced engineering plastics would be accessible to the Indian partner. The GE, on the other hand, found entry into the Indian market which was hitherto a highly protected one. PIL has been in this field using different raw materials and technology. The head of the unit is one of the directors of PIL and the head of the manufacturing has been recruited on the basis of his earlier experience in using a similar technology. The head of manufacturing and the quality assurance manager along with the other key personnel were given extensive training at the foreign partner's plant. The training combined both plant operations and courses in general management, marketing, and human resource development.

There is access to the available technology with the foreign partner, and the operations and quality departments in the new unit have been encouraged to bring in process modifications, product improvements, etc. This has resulted in major savings in terms of enhanced productivity, energy conservation, and product acceptability due to the innovative practices introduced in the new plant.

The manning norms have been decided with emphasis on flexibility on the job. There is a noticeable positive element in the blending of the Indian and GE culture to keep the manning levels to the absolute minimum. The personnel have been selected giving importance to their technical competence and their innovative skills. There is a considerable amount of flexibility built in each job. For example, the entire plant personnel are referred to as a manufacturing team and all operators/technicians handle both operations and maintenance jobs. This culture is in total contrast to PIL's work culture where operations and maintenance are treated as two distinct jobs. They have introduced the multiskilling concept in their plant.

The alliance has provided the Indian partner with the latest technology only while there are some restrictions imposed on the market, inputs, etc. The market for the products of the new plant is restricted to the Middle East, South East Asia, India and the neighbouring countries. The present process adopted also required almost 80 per cent of the raw material and components to be imported from the foreign partner. This is due to the nature of the technology introduced which requires totally new grades of raw material input available only with the foreign partner.

The new venture has been set up with a view to introducing a more informal work culture. Creativity and innovation are encouraged. The work practices reflect a wide range of cultural forces—attitudes of employees towards the new working style and transplanting does not take place in isolation from these larger social conditions. They have designed a common uniform for all employees as is one of the practices followed in the foreign partner's plant. They have also been promoting indigenous managerial capability in the plant to achieve the set performance standards.

The R&D centre, being set up at a different location to undertake application research, is headed by an expatriate. This centre plans to take up various product applications research, raw material alternatives, etc. This alliance of PIL has sought to overcome some of the technology-related problems by adopting more well-defined strategies. The level of absorption of technology has been quite high though the actual extent of assimilation is yet to be seen since this is a fairly new setup. There are no immediate plans for expansion as the market is yet to be explored and marketing strategies are yet to be evolved.

Case Study of ABC Company

ABC Company

ABC company is a diversified integrated corporation dealing with information technology, consumer products, health care, systems engineering, lighting, financial services, etc. This study is confined to the information technology division of the company. The information technology division of ABC company is the largest wing with a total sales turnover of about Rs. 3000 million (total sales turnover of the company is Rs. 6000 million) and a profit of Rs. 80 million in the year 1992-93.

The information technology division is divided into two wings—one dealing with computers and the other with systems. The technology division is considered to be one of the fastest growing software companies in India including exports to the tune of Rs. 340 million in 1992-93. The company holds a major market share offering the widest range of products in this line.

The company strategy is to invest in R&D apart from alliances with global leaders and technology providers like AT&T, Tandem, San Micro Systems, Intel, etc. ABC company ventured into the arena of information technology in the early eighties and gradually expanded its operations from services to technology and products. Initially, the company started with the import of hardware from Motorola with the software packages imported from INTEL. The company had progressed from merely adopting industry stardards to implementation of sophisticated networking systems. The primary objective, however, remained that they capture a market share and keep pace with the changes in technology. The focus has not been so much on technology development, innovation and sale of technical know-how, etc.

ABC company has positioned itself as a mediator between the various technology providers and technology packagers. They are not taking up basic research or applications research but are merely providing the intermediate support service for packaging and sale of these products. ABC company customizes or modifies the technology to suit the requirements of customers or end users and also provides the necessary support services. For example, they import around 50 per cent of the hardware components and manufacture the end products by integrating the system with suitable user packages. ABC company has entered into different types of arrangements to suit business needs like marketing alliances, technical know-how, providing qualified design engineers, etc. The marketing alliances have led to the benefits like distribution network and advantages of the economies of scale (mass production). The exchange of qualified personnel is in terms of the training and experience, sharing of the latest technological developments in the technology providers organization. This kind of arrangement begins on a project-tied basis to establish an initial business relationship which has eventually resulted in the development of mutual trust and further collaboration. For example, in one such arrangement, the company was able to negotiate for import of tools and basic technology for substantial value addition in its local unit. There has been a considerable buildup of the knowledge and skill base as a result of this exchange. The engineers trained in specific areas abroad return to the organization and in turn train the specific groups in usage of systems, packages, etc.

Human Resource Development in ABC Company

ABC company employs around 1700 employees, 80 per cent of whom are qualified technical personnel. The organization professes that it gives utmost importance to HRD and constant skill upgradation. It has developed a well-integrated system of internal and external training through both the Annual Performance Appraisal report and the assessment done by an external agency employed specifically for this purpose. The organization claims that because of its emphasis on HRD, the turnover of employees is around only 12 per cent as compared to the industry average turnover of 18 to 20 per cent. The organization believes in internal growth and career planning for its own personnel. Even in the case of new projects in diversified areas, the first priority is given to the placement of the existing employees with appropriate training rather than through external recruitment. This also has contributed to the lower turnover rate of persons in the organization.

The organization believes that technology can not only be purchased but it could be brought in and assimilated by the personnel who are sent abroad for training and who in turn also bring in the expertise to develop similar technology themselves. The organization has a well-developed system of job rotation in line with their interest, aptitude and career planning. The personnel are also rotated on projects abroad to provide opportunities to all for exposure to the learning from outside the organization. It is usually the practice to identify persons with at least 4-5 years' experience in the organization to be sent for training on projects abroad.

Research and Development in ABC Company

In ABC company, R&D is considered important in the process of technology absorption. The total expenditure on R&D during the year 1992-93 was to the extent of about Rs. 65 million, which is around 2 per cent of the total sales turnover, and was increased to Rs. 140 million in 1993-94 constituting more than 3 per cent of the sales turnover. The ABC company's R&D investment ranks amongst the top 50 R&D investments in India. The R&D investment is focussed on frontier areas of technology to enable the company to build up a creditable skill and knowledge base which has facilitated the company in entering into strategic alliances with large multinationals. The company's R&D has an inherent strength in the development of micro and mini computers in incorporating the latest technology. As a result of these efforts, the company was able to design and launch several new products, some specially designed for international markets. R&D in the company is given equal importance in relation to other departments. The strength of R&D professionals in the company was around 450 in 1994-95 as compared to 300 in 1993-94. They work in close collaboration with the marketing division and are also intensively involved right from the inception stage of negotiations with clients and partners in technology.

The information technology industry faces intense competition and technological obsolescence due to the rapid and phenomenal changes in information technology in recent years. The organization has therefore only striven so far to keep pace with the changes in technology. In the process, they have been importing technology for value addition and sale with not much accent on technology innovation or absorption. Assimilation has been only to the extent essential for upgrading or modifying the technology to suit customer specifications.

Factors that Help/Hinder Technology Assimilation in PIL and ABC

Assessment of Technology

For any assimilation of transfer of technology, the recipient organization should have the capability to assess appropriately the technology which they intend importing. This calls for a proper assessment of technology gaps through appropriate database and developing an institutional mechanism for such an assessment. This also entails upgrading the skill of human resources to analyse and upgrade information. In the initial stages the PIL did not possess sufficient technical competence to assess the technology being negotiated for. Over the period, PIL has gained sufficient experience in related fields to source and assess the technology to enable it to bargain for the best available technology. The earlier agreements were very biased towards the licensee and too many restrictive clauses were incorporated which were not conducive for upgradation of the technology.

ABC company follows a systematic procedure for continuous assessment and updating of the latest technological developments and market potential of the technology. They have been investing in upgrading human skills through research and training in frontier areas of technological development. Because of this constant updating of the investment in R&D, the company is in a better position to assess the global competitiveness of technology and also approach the technology developers awareness of latest market development, customer preferences, etc., which are necessary to keep pace with the changing technologies.

Internal Technology Development and Marketing Strategies

In PIL, the process of assimilation was not focussed on product upgradation and proper marketing strategies. The PIL operated in a protected market and this overlooked the need for forecasting technology development and suitably upgrading their skills. Now the opening up of the economy and entry of new industries have put tremendous pressure on exploring new markets. ABC company operates in a highly competitive market environment which has necessitated the adoption of aggressive and innovative strategies to retain and expand its market. There is a constant upgrading of technology and product features to suit customer needs and specifications with a backup of support services to enhance the marketability of its products.

Negotiating for Technology Alliance

The choice, evaluation and negotiation mechanism are very crucial for any technology alliance. The team constituted in PIL for negotiating for technology is based on expertise. At times, there are pressures from the government or the top management of the company to send persons on expediency rather than expertise. There are also limitations imposed regarding the number of persons who could be sent for negotiations. For instance, in a recent negotiation with a US based licensor, the negotiating team faced major problems in obtaining clearances for visit to the US. The government was willing to allow only two representatives— and insisted on the Director being one. After a lot of follow-up, the three-member team was given the permission. During the course of negotiations, the US company tried incorporating the restrictive 302 clauses in the agreement. It was the presence of the technical person which helped in avoiding this clause being included as he was successful in arguing against validity and anti-trust clause. It was offered that PIL could obtain the technology know-how documents through the other licensee of the American firm located in Asia if it were possible. Earlier trends of PIL showed that they had scrupulously observed the restrictive clauses imposed by licensee.

In the earlier years, in PIL, the emphasis was on getting readymade technology without real attention being given to internal assimilation and upgradation. In many cases, due to the weak negotiating position, such upgradation was also restricted. Because of the monopoly enjoyed by PIL, there was little incentive to upgrade the technology even after the expiry of the agreement and restriction period. It is only recently that with competition increasing, the company is now introducing CAD and CAM technologies for simpler controls and better documentation and data retrieval systems.

For any assimilation and upgradation of technology the understanding of the process aspect of technology is important. However, in PIL initially due to restrictive clauses and later due to lack of initiative the transfer of technology was confined to operational level. Even the training given to the employees was confined to the operational aspects. Since the R&D in PIL has a limited role, for any upgradation, the company repeatedly went in for further technical collaboration paying professional charges.

ABC company builds the team for negotiations for alliances with persons having the best expertise, competence and experience and position is not the only criterion for selection. The process of negotiation for any tie-up or alliance is initiated very gradually after building up a rapport and goodwill with the other organization. For example, the ABC company first provided only design experts to a US-based company to help them in designing and programming. Over a period of two years, after establishing a definite level of confidence and a positive working relationship, ABC company approached the other organization for an alliance to operate from India.

It is worth noting that the constitution of the negotiation team and process are at times constrained by rules and regulations in the public sector; the private sector has the freedom to select the best possible teams based on their expertise with the result that they are able to take faster decisions and get better terms.

Issues of Assimilation and Incentives

Though PIL has been successful in developing and commercializing certain technologies, it has not been successful in marketing its technology know-how. It has been PIL's experience that the Indian market is biased towards foreign tie-ups for the brand name even though indigenous technology is available—both proven and cheaper technology. Also, the company has not been aggressive in selling its technology.

The nature of technology and products dealt with by ABC company, is such that requires a high level of assimilation and upgradation of technology. The products and processes are imported with a view to marketing them after suitable value addition and customizing products to the extent required by the clients. This in itself has been and an incentive for constant upgradation of technology. The access to technology has been through training in organizations abroad and continuous involvement of R&D in all areas of business activity on better terms and conditions for technology transfer.

Interdepartmental Coordination and Role of R&D

Another noticeable feature in PIL was that even after the transfer of technology, the interaction between R&D and operations was not intensive and there was a noticeable lack of coordination between them. There was a reluctance in the sharing of information and documentation between the two departments.

PIL has a technology department which acts as a coordinating agency between the R&D and operations department. The role of technology department includes monitoring the production process, identifying areas for improvement, and involving R&D in bringing necessary changes or modifications. The role of R&D is, therefore, restricted to some troubleshooting, attempting indigenization of certain raw materials, enhancing/optimizing consumption levels, etc.

The role of the technology department could be more intensive at the time of technology transfer and gradually R&D department could play a greater role along with the operations in order to constantly improve and upgrade technology through closer monitoring. The technology and R&D department could be merged so that the R&D does not operate in isolation of operations and also take up more applied research activities.

One noticeable feature in ABC company was the role of R&D in the mainstream activity and its involvement in all the major activities including negotiations. Marketing and R&D are treated at par and work in very close coordination. R&D is very actively involved in technology development and applications research to improve product lines and features.

Marketing of Technology Capability

The R&D in PIL has so far made a few significant contributions in terms of developing certain catalysts and process modifications. PIL is now able to offer certain advances and improvements in the technology and also show greater potential for assimilation. This has helped in the alliance with larger firms due to the technical backup and R&D inputs available within the organization.

ABC company has established a reputation of truthworthiness and quality in the global R&D and marketing capabilities. Technology, products and systems developed by ABC company have been accepted in the international and domestic markets. (e.g. UNIX operating system)

Human Resource Related Issues

Another factor hindering the proper assimilation process is the manning pattern which got evolved in PIL over the years. In one of the studies conducted in the organization, it was found that around 30 per cent of employees both at managerial and workmen level were surplus and also a number of them had overlapping responsibilities. During the project implementation stage, proper caution is not observed in matching the persons and the skills required. Further identification of personnel transfers and decisions is delayed resulting in limited time being available for adequate preparation for the new job. The selection of persons for training, specially abroad, is based on seniority principle, rather than on the competence of persons to handle the job or utilize the learning after training on the job. For instance, one of the senior managers in the personnel department was sent for training on advanced welding technology with the collaborators abroad simply because he was the seniormost person and it was his turn based on his seniority to undergo technical training. After his return, he was placed back into the personnel department and within a year of his return he retired from service. The entire training effort was therefore a waste. There were many such instances quoted in the organization, where training was not properly utilized. This indicates a lack of career planning, training and placement policies in the organization.

While selecting people for handling new projects, the criteria applied do not necessarily include previous training and experience in the particular line. This is partly due to the organizational culture to give a greater number of persons the opportunity to handle new projects and undergo training which results in the organization not utilizing the best available skills and expertise.

Trade unions and officer associations play a dominant role in slowing down the proper assimilation and upgradation of technology by offering overt pressure for bringing in change. In PIL, due to technology change and minor process modification, no redundancies are allowed to be declared by trade unions. This hampers upgradation and technology change. This indicates a lack of appropriate redeployment policy on the part of organization.

ABC company believes in the proper training and development of personnel. The major problem remains the high turnover which, although lower than the industry average in ABC company, does present certain constraints. The career planning and growth paths are reviewed periodically to overcome the problems in succession planning. There is a systematic training need identification process to help individuals and organization upgrade skills. The selection process for initial recruitment tries to maintain a balance between experience and newly acquired skills. With the opening up of the economy and the entry of multinationals into the market, and shortage of skills, the salary and compensation are very high to attract the best. ABC company also has had to increase its incentives in pay packages adding to the overall manpower costs considerably.

Manufacturing Activities

The liberalization and changes in customs and import duties have drastically affected the manufacturing operations in the information technology industry. ABC company has substantially reduced its own basic manufacturing activity as importing of the bulk of components has become simpler and cheaper and to take advantage of the economies of scale afforded by larger manufacturers (e.g. software division). This has affected technology assimilation to an extent.

Conclusions/Suggestions

(1) ***Technology Improvement and Image Building:*** In Indian organizations, in a substantial number of cases the incentive for technology alliances/transfer is to get a foreign brand name which in India has a better saleability. This always is to the disadvantage of the Indian partner, specially in case where technologically the Indian organizations are weak and the major incentive offered is the vast marketing potential. To overcome this problem gradually, the organization should develop long-term and short-term technology assimilation policies which eventually will help in image building of the organization.

(2) ***Buliding and Marketing of Indian Organizational Image :*** It is found that in some organizations they do have tremendous capability and technical competence to develop and innovate technology as was found in some areas in the PIL case. However, the selling of the competence is weak and the organization must make a conscious effort at building their image with regard to their present and potential technology capability. ABC company has conciously established an image and concentrate on developing and maintaining an image of quality and service.

(3) ***Incentives for Technicians***: In an Indian organization, the employee growth prospects are always higher in administrative responsibilities and jobs. Therefore even the capable technicians normally aspire for

administrative work, due to lack of career prospects in their own technical areas. In contrast, in Korea, the technician can rise up to the master craftsman level, by passing certain examinations and meeting laid-down criteria. According to Korean law, a master craftsman has to be paid a salary equivalent to that of a full professor. This has enabled the specialized skills to be retained and improved within the organization/country. It has also contributed to the reversal of brain-drain.

(4) ***Increasing R&D Involvement:*** The R&D involvement in technology transfer and assimilation in India is weak. Even the R&D budgets are not sufficient and in many cases, these are there only to avail tax benefits. Even in ABC company, though R&D is given importance, it is mainly to keep pace with the latest developments and not for innovations and changes. In any technology transfer as well as assimilation the R&D personnel should be involved right from the beginning and they may be given the major responsibility of assimilation and upgradation.

(5) ***Documentation:*** For any transfer and assimilation of technology, the systematic documentation of the entire process of negotiation, transfer and technical details including any upgradation and modification from time to time is essential. Proper documentation of all relevant information enables simpler and quicker information sharing and review of the level of technological development and also ensures continuity of development efforts even in case of high employee turnover.

(6) ***Closer Collaboration between Government Research & Development and Industry:*** As the financial resources for setting up R&D centres is limited, there could be much closer collaborations between government R&D laboratories (CSIR) and industry, specially to take up joint applied research projects. Even some of the universities or technical institutions like the Indian Institutes of Technology could also collaborate with Indian R&D labs by constituting research teams for specific projects.

(7) ***Increasing the Trainability of Workforce:*** The Human Resource Development aspects, specially of the workmen is haphazard and weak. More investment needs to be made in skill upgradation as well as increasing the

trainability of the employees specially in sectors where technology is fast changing— like electronics. Some companies like ABC are investing in training mainly to meet their immediate needs while there is very little investment for future skill development.

(8) ***Negotiation Capabilities:*** Many Indian companies do not negotiate the technology transfer for the purpose of assimilation; therefore, it is transfer of packaged technology. With the vast availability of qualified human resource in India, it could be possible to gradually develop one's own capability for equitable alliances rather than a pure technology alliance.

(9) ***Blending of Technology and Adopting Indigenous Norms:*** At present, there is a wholesale transfer of technology including the manning norms and production yardsticks. Due to deficiencies in skill level of employees, the educational background, climatic conditions, etc., in most cases those norms become inoperable. There is need to do an appropriate blending of technology and at the same time develop different norms and production yardsticks. For example, the manning norms and productivity yardsticks could be based on investment per man or production per unit of capital employed rather than on man-hours.

Another related issue is the employment and development of the skilled professionals. Indian organizations should not only consider their operating capabilities, but also their cognitive abilities for assimilation of technology.

(10) ***Role of Government Policies:*** In the past, the government regulations restricted not only transfer but also assimilation of technology by enforcing too many regulatory norms. With recent liberalization measures, these regulations are being relaxed. The role of government policies should be more facilitative and less regulatory like for example in Malaysia and Thailand. These governments have taken the responsibility of providing infrastructual facilities, training the workforce, bringing in a more conducive legal framework for regulating employer/employee relations and at the same time protecting the employees' interests. A similar approach in India could help the assimilation process.

(11) ***Cultural Factors in Transfer of Technology:*** While evolving the Human Resources policies, given the limitations of recruitment, promotion, etc., more emphasis has to be given to training as compared to some of the advanced countries. Similarly, the annual appraisal system could be used as a tool for development rather than as a tool for reward and punishment.

(12) ***Incentives for Innovation:*** At the firm level, further encouragement and incentives for assimilation and innovation should be introduced to enhance the process of assimilation. As is seen in the latest alliance of PIL, where this is being consciously rewarded with almost immediate effect, there have been major contributions from the operating staff rather than the research labs in terms of process modifications, better and simpler control systems, etc.

(13) ***Management of R&D for Industry-Academy Collaboration:*** In many countries, it is found that though heavy investment is made in technical education, equipment, infrastructure and modern computers, yet there is hardly any adequate return in terms of development of applied research projects by which a country can benefit. In many cases, it is also found that the faculty of higher technical institutions are inclined to do basic research which helps them in publishing papers and enhancing their academic stature. However, this has not been usefully adopted by business and industry in the areas of applied research as they do not use sophisticated methodology and to that extent are not generally favoured by the faculty with the result though there may be highly sophisticated institutions of technical education yet their spin-off benefits to the surrounding environment is limited. They remain isolated packages of excellence without much direct interaction with the neighbourhood business and industry. Though these institutions are well-equipped with all the infrastructure facilities yet no space is earmarked for research environment where academics can share information with representatives from industrial houses and business enterprises. On the other hand, the business enterprises lack the necessary infrastructure and facilities needed for research. Therefore, some of the organizations and institutions may consider establishing 'research parks' with the following guidelines:

1. Space in the institutions of excellence should be made available to industrial houses and business enterprises on a lease basis with facilities for research.

2. Such facilities should form the nucleus of research parks with the most modern amenities and facilities for research.

3. These research parks should have a standing relationship with academicians for sharing of applied research results.

4. Research parks may be controlled by specially constituted committees appointed by the governing councils of various institutions and should have representatives from several industries.

The expenses for establishing these research parks should be met out of the lease money on a time-bound basis and from other sources like facilities provided to the industrial houses. These research parks should be exclusively meant for applied research projects and should have no commercial objectives. These projects should have some relationship with the capabilities and potential of the concerned institution. The projects should utilize directly or otherwise the faculties, libraries, equipment and other infrastructure in the institution. Some balancing arrangement of equipment and instruments for research should be made in the research parks. Industries and enterprises could have full rights to the end-products of such a project. However, the methodology involved for research may be used by academicians for teaching and publication purposes. Methodology and data collection could be shared with other institutions. Such research parks have been established in the USA in as many as 80 institutions.

Summing Up

Any HRD planning process must take into account the level of development of each country. Therefore, there cannot be uniform HRD strategies applicable to all situations. Even within the country, there could be wide variations in terms of the development of technology in different segments and sectors of economy. Though it is often argued that new markets and technologies require a more highly skilled, better educated, more mobile workforce, yet, on the other hand, in many developing countries, such statements could be contradicted based on employment level and opportunities available. There may be a shift from manual jobs to more white-collar jobs. However, many workers are still struggling in work environment which has not altered since the middle of the last century. Many more workers are caught up in the movement towards greater automation and at times perform mindless jobs; at the same time we may have some kind of technology importation. Therefore, it may be impractical to assume that there will be a sudden change of technology in all spheres in all countries. In certain segments and sectors, there may be modern technology, while in other spheres, especially in the informal sector, traditional technology with slight improvements may continue. Therefore, the countries have to develop people with different training or different skills to suit the traditional, most modern and blended version of the two.

BIBLIOGRAPHY

US-Japan Strategic Alliances in the Semiconductor Industry, Committee on Japan, Technology Transfer, Competition, and Public Policy, National Research Council, Washington D.C., 1992.

Technology Atlas—An Interim Report Asia and Pacific Centre for Transfer of Technology of the ESCAP, 1987.

Indian Management Virmani B.R. & Suynil Unny Guptan, (Vision Books, New Delhi), 1991.

Dr. Peter Lorange, Strategic Alliances : Formation, Implementation and Evolution, Interman Reports for the Business Community, (International Labour Office, Geneva 22, Switzterland.

Charles T. Stewart. Jr. and Yasumitsu Niheli, *Technology Transfer and Human Factors* Lexington Press, 1992.

Virmani B.R.,*Corporate Environmentalism in a Global Economy—Societal Values in International Technology Transfer*, Virmani B.R., Quorum Books, Westport Connecticut, London, 1993.

Technology Transfer, Assimilation and Human Resource Management, Virmani B.R. & Kala Rao, *Personnel Today*, July-September, 1994.

Economic Restructuring, Technology Transfer and Human Resource Development, Sage Publications, New Delhi, 1996 (under publication).

